Developing
Essential Understanding
of
Rational Numbers
for Teaching Mathematics *in*
Grades 3–5

Carne Barnett-Clarke
WestEd
Oakland, California

William Fisher
California State University, Chico
Chico, California

Randall I. Charles
Volume Editor
Carmel, California

Rick Marks
Sonoma State University
Rohnert Park, California

Rose Mary Zbiek
Series Editor
The Pennsylvania State University
University Park, Pennsylvania

Sharon Ross
California State University, Chico
Chico, California

NATIONAL COUNCIL OF
TEACHERS OF MATHEMATICS
NCTM®

D1218674

Copyright © 2010 by
The National Council of Teachers of Mathematics, Inc.
1906 Association Drive, Reston, VA 20191-1502
(703) 620-9840; (800) 235-7566; www.nctm.org
All rights reserved
Fourth printing 2016

Library of Congress Cataloging-in-Publication Data

Developing essential understanding of rational numbers for teaching
mathematics in grades 3/5 / Carne Barnett-Clarke ... [et al.].
    p. cm.
  Includes bibliographical references.
  ISBN 978-0-87353-630-1
  1. Numbers, Rational—Study and teaching (Elementary) 2.
Mathematics—Study and teaching (Elementary) I. Barnett-Clarke, Carne.
  QA135.6.D487 2010
  372.7--dc22

                                        2010033718

The National Council of Teachers of Mathematics is the public voice of mathematics education, supporting teachers to ensure equitable mathematics learning of the highest quality
for all students through vision, leadership, professional development, and research.

Printed in the United States of America

Contents

Foreword v

Preface vii

Introduction 1
  Why Rational Numbers? 1
  Understanding Rational Numbers 2
  The Big Ideas and Essential Understandings............................ 3
  Benefits for Teaching, Learning, and Assessing....................... 4
  Ready to Begin 5

Chapter 1 7
Rational Numbers: The Big Ideas and Essential Understandings

  Extending Our Use of Numbers: Big Idea 1............................. 10
    From counting to measurement 11
    Defining *rational number*... 13
    Finding closure with rational numbers.............................. 17

  Making Sense of Rational Numbers: Big Idea 2........................ 19
    Unit as the basis for interpretation............................. 19
    Multiple interpretations: Rational numbers as part-whole
      relationships 20
    Multiple interpretations: Rational numbers as measures............ 21
    Multiple interpretations: Rational numbers as quotients........... 24
    Multiple interpretations: Rational numbers as ratios.............. 25
    Multiple interpretations: Rational numbers as operators........... 27
    An expanded view of *unit*.. 28

  Rational Numbers and Equivalence: Big Idea 3........................ 30
    Representing rational numbers as equivalent fractions............. 30
    Rational numbers and density 34
    Representing rational numbers in decimal form 36

  Computing with Rational Numbers: Big Idea 4 42
    Interpreting rational number operations and algorithms........... 42
    Estimation and mental math with rational numbers................. 55

  Conclusion.. 56

Chapter 2 59

Connections: Looking Back and Ahead in Learning

Building from Whole Numbers to Rational Numbers 59
Counting, unitizing, and multiplication 59
Fractions and division 60
Decimals and place value 60

Using Rational Numbers across the Curriculum 61
Measurement 61
Ratios and proportional reasoning 61
Percents 62
Probability and data analysis 63
Algebra 64

Extending beyond Rational Numbers 66
Negatives of fractions 66
Irrational and real numbers 66
Other numbers 67

Conclusion 68

Chapter 3 69

Challenges: Learning, Teaching, and Assessing

Relating Rational Numbers to Whole Numbers 69
Shift 1–From unrelated system to natural extension 70
Facilitating Shift 1 70

Multiple Interpretations of Rational Numbers and Units 72
Shift 2–From one model to many representations 73
Facilitating Shift 2 74

Understanding Rational Number Equivalence 75
Shift 3–From whole number–based to equivalence-based comparisons 76
Facilitating Shift 3 76

Making Sense of Rational Number Operations 77
Shift 4–From "rules" to sense making 78
Facilitating Shift 4 78

Conclusion 80

References 81

Foreword

Teaching mathematics in prekindergarten–grade 12 requires a special understanding of mathematics. Effective teachers of mathematics think about and beyond the content that they teach, seeking explanations and making connections to other topics, both inside and outside mathematics. Students meet curriculum and achievement expectations when they work with teachers who know what mathematics is important for each topic that they teach.

The National Council of Teachers of Mathematics (NCTM) presents the Essential Understanding Series in tandem with a call to focus the school mathematics curriculum in the spirit of *Curriculum Focal Points for Prekindergarten through Grade 8 Mathematics: A Quest for Coherence*, published in 2006, and *Focus in High School Mathematics: Reasoning and Sense Making,* released in 2009. The Essential Understanding books are a resource for individual teachers and groups of colleagues interested in engaging in mathematical thinking to enrich and extend their own knowledge of particular mathematics topics in ways that benefit their work with students. The topic of each book is an area of mathematics that is difficult for students to learn, challenging to teach, and critical for students' success as learners and in their future lives and careers.

Drawing on their experiences as teachers, researchers, and mathematicians, the authors have identified the big ideas that are at the heart of each book's topic. A set of essential understandings—mathematical points that capture the essence of the topic—fleshes out each big idea. Taken collectively, the big ideas and essential understandings give a view of a mathematics that is focused, connected, and useful to teachers. Links to topics that students encounter earlier and later in school mathematics and to instruction and assessment practices illustrate the relevance and importance of a teacher's essential understanding of mathematics.

On behalf of the Board of Directors, I offer sincere thanks and appreciation to everyone who has helped to make this series possible. I extend special thanks to Rose Mary Zbiek for her leadership as series editor. I join the Essential Understanding project team in welcoming you to these books and in wishing you many years of continued enjoyment of learning and teaching mathematics.

Henry Kepner
President, 2008–2010
National Council of Teachers of Mathematics

Preface

From prekindergarten through grade 12, the school mathematics curriculum includes important topics that are pivotal in students' development. Students who understand these ideas cross smoothly into new mathematical terrain and continue moving forward with assurance.

However, many of these topics have traditionally been challenging to teach as well as learn, and they often prove to be barriers rather than gateways to students' progress. Students who fail to get a solid grounding in them frequently lose momentum and struggle in subsequent work in mathematics and related disciplines.

The Essential Understanding Series identifies such topics at all levels. Teachers who engage students in these topics play critical roles in students' mathematical achievement. Each volume in the series invites teachers who aim to be not just proficient but outstanding in the classroom—teachers like you—to enrich their understanding of one or more of these topics to ensure students' continued development in mathematics.

How much do you need to know?

To teach these challenging topics effectively, you must draw on a mathematical understanding that is both broad and deep. The challenge is to know considerably more about the topic than you expect your students to know and learn.

Why does your knowledge need to be so extensive? Why must it go above and beyond what you need to teach and your students need to learn? The answer to this question has many parts.

To plan successful learning experiences, you need to understand different models and representations and, in some cases, emerging technologies as you evaluate curriculum materials and create lessons. As you choose and implement learning tasks, you need to know what to emphasize and why those ideas are mathematically important.

While engaging your students in lessons, you must anticipate their perplexities, help them avoid known pitfalls, and recognize and dispel misconceptions. You need to capitalize on unexpected classroom opportunities to make connections among mathematical ideas. If assessment shows that students have not understood the material adequately, you need to know how to address weaknesses that you have identified in their understanding. Your understanding must be sufficiently versatile to allow you to represent the mathematics in different ways to students who don't understand it the first time.

In addition, you need to know where the topic fits in the full span of the mathematics curriculum. You must understand where your students are coming from in their thinking and where they are heading mathematically in the months and years to come.

Accomplishing these tasks in mathematically sound ways is a tall order. A rich understanding of the mathematics supports the varied work of teaching as you guide your students and keep their learning on track.

How can the Essential Understanding Series help?

The Essential Understanding books offer you an opportunity to delve into the mathematics that you teach and reinforce your content knowledge. They do not include materials for you to use directly with your students, nor do they discuss classroom management, teaching styles, or assessment techniques. Instead, these books focus squarely on issues of mathematical content—the ideas and understanding that you must bring to your preparation, in-class instruction, one-on-one interactions with students, and assessment.

How do the authors approach the topics?

For each topic, the authors identify "big ideas" and "essential understandings." The big ideas are mathematical statements of overarching concepts that are central to a mathematical topic and link numerous smaller mathematical ideas into coherent wholes. The books call the smaller, more concrete ideas that are associated with each big idea *essential understandings*. They capture aspects of the corresponding big idea and provide evidence of its richness.

The big ideas have tremendous value in mathematics. You can gain an appreciation of the power and worth of these densely packed statements through persistent work with the interrelated essential understandings. Grasping these multiple smaller concepts and through them gaining access to the big ideas can greatly increase your intellectual assets and classroom possibilities.

In your work with mathematical ideas in your role as a teacher, you have probably observed that the essential understandings are often at the heart of the understanding that you need for presenting one of these challenging topics to students. Knowing these ideas very well is critical because they are the mathematical pieces that connect to form each big idea.

How are the books organized?

Every book in the Essential Understanding Series has the same structure:

- The introduction gives an overview, explaining the reasons for the selection of the particular topic and highlighting some of the differences between what teachers and students need to know about it.

Big ideas and essential understandings are identified by icons in the books.

marks a big idea,

and

marks an essential understanding.

- Chapter 1 is the heart of the book, identifying and examining the big ideas and related essential understandings.

- Chapter 2 reconsiders the ideas discussed in chapter 1 in light of their connections with mathematical ideas within the grade band and with other mathematics that the students have encountered earlier or will encounter later in their study of mathematics.

- Chapter 3 wraps up the discussion by considering the challenges that students often face in grasping the necessary concepts related to the topic under discussion. It analyzes the development of their thinking and offers guidance for presenting ideas to them and assessing their understanding.

The discussion of big ideas and essential understandings in chapter 1 is interspersed with questions labeled "Reflect." It is important to pause in your reading to think about these on your own or discuss them with your colleagues. By engaging with the material in this way, you can make the experience of reading the book participatory, interactive, and dynamic.

Reflect questions can also serve as topics of conversation among local groups of teachers or teachers connected electronically in school districts or even between states. Thus, the Reflect items can extend the possibilities for using the books as tools for formal or informal experiences for in-service and preservice teachers, individually or in groups, in or beyond college or university classes.

marks a "Reflect" question that appears on a different page.

A new perspective

The Essential Understanding Series thus is intended to support you in gaining a deep and broad understanding of mathematics that can benefit your students in many ways. Considering connections between the mathematics under discussion and other mathematics that students encounter earlier and later in the curriculum gives the books unusual depth as well as insight into vertical articulation in school mathematics.

The series appears against the backdrop of *Principles and Standards for School Mathematics* (NCTM 2000), *Curriculum Focal Points for Prekindergarten through Grade 8 Mathematics: A Quest for Coherence* (NCTM 2006), *Focus in High School Mathematics: Reasoning and Sense Making* (NCTM 2009), and the Navigations Series (NCTM 2001–2009). The new books play an important role, supporting the work of these publications by offering content-based professional development.

The other publications, in turn, can flesh out and enrich the new books. After reading this book, for example, you might select hands-on, Standards-based activities from the Navigations books for your students to use to gain insights into the topics that the

Essential Understanding books discuss. If you are teaching students in prekindergarten through grade 8, you might apply your deeper understanding as you present material related to the three focal points that *Curriculum Focal Points* identifies for instruction at your students' level. Or if you are teaching students in grades 9–12, you might use your understanding to enrich the ways in which you can engage students in mathematical reasoning and sense making as presented in *Focus in High School Mathematics*.

An enriched understanding can give you a fresh perspective and infuse new energy into your teaching. We hope that the understanding that you acquire from reading the book will support your efforts as you help your students grasp the ideas that will ensure their mathematical success.

We would like to give particular thanks to the following individuals, who reviewed an earlier version of this book: Dale R. Oliver, Shari Reed, Tad Watanabe, and Jennifer Zahuranec. Their insights, as well as our conversations with many other colleagues, inspired our ideas and challenged our thinking.

Introduction

This book focuses on ideas about rational numbers. These are ideas that you need to understand thoroughly and be able to use flexibly to be highly effective in your teaching of mathematics in grades 3–5. The book discusses many mathematical ideas that are common in elementary school curricula, and it assumes that you have had a variety of mathematics experiences that have motivated you to delve into—and move beyond—the mathematics that you expect your students to learn.

The book is designed to engage you with these ideas, helping you to develop an understanding that will guide you in planning and implementing lessons and assessing your students' learning in ways that reflect the full complexity of rational numbers. A deep, rich understanding of ideas about rational numbers will enable you to communicate their influence and scope to your students, showing them how these ideas permeate the mathematics that they have encountered—and will continue to encounter—throughout their school mathematics experiences.

The understanding of rational numbers that you gain from this focused study thus supports the vision of *Principles and Standards for School Mathematics* (NCTM 2000): "Imagine a classroom, a school, or a school district where all students have access to high-quality, engaging mathematics instruction" (p. 3). This vision depends on classroom teachers who "are continually growing as professionals" (p. 3) and routinely engage their students in meaningful experiences that help them learn mathematics with understanding.

Why Rational Numbers?

Like the topics of all the volumes in the Essential Understanding Series, rational numbers compose a major area of school mathematics that is crucial for students to learn but challenging for teachers to teach. Students in grades 3–5 need to understand rational numbers well if they are to succeed in these grades and in their subsequent mathematics experiences. Learners often struggle with ideas about rational numbers. What is the relationship between rational numbers and fractions, for example? Many students mistakenly believe that these are identical. The importance of rational numbers and the challenge of understanding them well make them essential for teachers of grades 3–5 to understand extremely well themselves.

Your work as a teacher of mathematics in these grades calls for a solid understanding of the mathematics that you—and your school, your district, and your state curriculum—expect your students to learn about rational numbers. Your work also requires you to know how this mathematics relates to other mathematical ideas that your students will encounter in the lesson at hand, the current school year, and beyond. Rich mathematical understanding guides teachers' decisions in much of their work, such as choosing tasks for a lesson, posing questions, selecting materials, ordering topics and ideas over time, assessing the quality of students' work, and devising ways to challenge and support their thinking.

Understanding Rational Numbers

Teachers teach mathematics because they want others to understand it in ways that will contribute to success and satisfaction in school, work, and life. Helping your students develop a robust and lasting understanding of rational numbers requires that you understand this mathematics deeply. But what does this mean?

It is easy to think that understanding an area of mathematics, such as rational numbers, means knowing certain facts, being able to solve particular types of problems, and mastering relevant vocabulary. For example, for the upper elementary grades, you are expected to know facts such as "the set of whole numbers is a subset of the rational numbers." You are expected to be skillful in solving problems that involve multiplying fractions. Your mathematical vocabulary is assumed to include such terms as *fraction*, *rational number*, *reciprocal*, and *lowest common denominator*.

Obviously, facts, vocabulary, and techniques for solving certain types of problems are not all that you are expected to know about rational numbers. For example, in your ongoing work with students, you have undoubtedly discovered that you need to distinguish among different interpretations of rational numbers, such as knowing the difference between rational numbers used as measures and rational numbers used as quotients.

It is also easy to focus on a very long list of mathematical ideas that all teachers of mathematics in grades 3–5 are expected to know and teach about rational numbers. Curriculum developers often devise and publish such lists. However important the individual items might be, these lists cannot capture the essence of a rich understanding of the topic. Understanding rational numbers deeply requires you not only to know important mathematical ideas but also to recognize how these ideas relate to one another. Your understanding continues to grow with experience and as a result of opportunities to embrace new ideas and find new connections among familiar ones.

Furthermore, your understanding of rational numbers should transcend the content intended for your students. Some of the differences between what you need to know and what you expect them to learn are easy to point out. For instance, your understanding of the topic should include a grasp of the way in which rational numbers connect with irrational numbers—mathematics that students will encounter later but do not yet understand.

Other differences between the understanding that you need to have and the understanding that you expect your students to acquire are less obvious, but your experiences in the classroom have undoubtedly made you aware of them at some level. For example, how many times have you been grateful to have an understanding of rational numbers that enables you to recognize the merit in a student's unanticipated mathematical question or claim? How many other times have you wondered whether you could be missing such an opportunity or failing to use it to full advantage because of a gap in your knowledge?

As you have almost certainly discovered, knowing and being able to do familiar mathematics are not enough when you're in the classroom. You also need to be able to identify and justify or refute novel claims. These claims and justifications might draw on ideas or techniques that are beyond the mathematical experiences of your students and current curricular expectations for them. For example, you may need to be able to refute the often-asserted, erroneous claim that all ratios are fractions. Or you may need to explain to a student why the product of two fractions between 0 and 1 is less than either factor.

The Big Ideas and Essential Understandings

Thinking about the many particular ideas that are part of a rich understanding of rational numbers can be an overwhelming task. Articulating all of those mathematical ideas and their connections would require many books. To choose which ideas to include in this book, the authors considered a critical question: What is *essential* for teachers of mathematics in grades 3–5 to know about rational numbers to be effective in the classroom? To answer this question, the authors drew on a variety of resources, including personal experiences, the expertise of colleagues in mathematics and mathematics education, and the reactions of reviewers and professional development providers, as well as ideas from curricular materials and research on mathematics learning and teaching.

As a result, the mathematical content of this book focuses on essential ideas for teachers about rational numbers. In particular, chapter 1 is organized around four big ideas related to this important area of mathematics. Each big idea is supported by smaller, more specific mathematical ideas, which the book calls *essential understandings*.

Benefits for Teaching, Learning, and Assessing

Understanding rational numbers can help you implement the Teaching Principle enunciated in *Principles and Standards for School Mathematics*. This Principle sets a high standard for instruction: "Effective mathematics teaching requires understanding what students know and need to learn and then challenging and supporting them to learn it well" (NCTM 2000, p. 16). As in teaching about other critical topics in mathematics, teaching about rational numbers requires knowledge that goes "beyond what most teachers experience in standard preservice mathematics courses" (p. 17).

Chapter 1 comes into play at this point, offering an overview of rational numbers that is intended to be more focused and comprehensive than many discussions of the topic that you are likely to have encountered. This chapter enumerates, expands on, and gives examples of the big ideas and essential understandings related to rational numbers, with the goal of supplementing or reinforcing your understanding. Thus, chapter 1 aims to prepare you to implement the Teaching Principle fully as you provide the support and challenge that your students need for robust learning about rational numbers.

Consolidating your understanding in this way also prepares you to implement the Learning Principle outlined in *Principles and Standards*: "Students must learn mathematics with understanding, actively building new knowledge from experience and prior knowledge" (NCTM 2000, p. 20). To support your efforts to help your students learn about rational numbers in this way, chapter 2 builds on the understanding of rational numbers that chapter 1 communicates by pointing out specific ways in which the big ideas and essential understandings connect with mathematics that students typically encounter earlier or later in school. This chapter supports the Learning Principle by emphasizing longitudinal connections in students' learning about rational numbers. For example, as their mathematical experiences expand, students gradually develop an understanding of the connections between fractions and decimals and become fluent in using equivalent representations as needed.

The understanding that chapters 1 and 2 convey can strengthen another critical area of teaching. Chapter 3 addresses this area, building on the first two chapters to show how an understanding of rational numbers can help you select and develop appropriate tasks, techniques, and tools for assessing your students' understanding of rational numbers. An ownership of the big ideas and essential understandings related to rational numbers, reinforced by an understanding of students' past and future experiences with the ideas, can help you ensure that assessment in your classroom supports the learning of significant mathematics.

Such assessment satisfies the first requirement of the Assessment Principle set out in *Principles and Standards*: "Assessment should support the learning of important mathematics and furnish useful information to both teachers and students" (NCTM 2000, p. 22). An understanding of rational numbers can also help you satisfy the second requirement of the Assessment Principle, by enabling you to develop assessment tasks that give you specific information about what your students are thinking and what they understand. For example, the following task challenges students to think about the idea that the quantity represented by a fraction is relative to the size of the unit:

Consider line segment *AB* below:

$$A \hspace{6cm} B$$

Draw a line segment that is 1 unit long if line segment *AB* represents—

$$a.\ \frac{1}{3}\ \text{unit} \qquad b.\ \frac{2}{3}\ \text{unit} \qquad c.\ 1\frac{1}{2}\ \text{units}$$

Ready to Begin

This introduction has painted the background, preparing you for the big ideas and associated essential understandings related to rational numbers that you will encounter and explore in chapter 1. Reading the chapters in the order in which they appear can be a very useful way to approach the book. Read chapter 1 in more than one sitting, allowing time for reflection. Absorb the ideas—both big ideas and essential understandings—related to rational numbers. Appreciate the connections among these ideas. Carry your newfound or reinforced understanding to chapter 2, which guides you in seeing how the ideas related to rational numbers are connected to the mathematics that your students have encountered earlier or will encounter later in school. Then read about teaching, learning, and assessment issues in chapter 3.

Alternatively, you may want to take a look at chapter 3 before engaging with the mathematical ideas in chapters 1 and 2. Having the challenges of teaching, learning, and assessment issues clearly in mind, along with possible approaches to them, can give you a different perspective on the material in the earlier chapters.

No matter how you read the book, let it serve as a tool to expand your understanding, application, and enjoyment of rational numbers.

Chapter 1

Rational Numbers: The Big Ideas and Essential Understandings

Why devote a book to rational numbers? The curriculum devotes a lot of time to work with fractions and decimals, yet both teachers and students commonly find the ideas and skills related to these numbers difficult. This book explores the mathematics of rational numbers in both fractional and decimal forms and their relationship to percents. It focuses particularly on fractions and emphasizes ideas that are important for teachers in grades 3–5 to understand, while tracing how rational numbers developed, what they mean, how they behave, and how we use them in our everyday world.

Four big ideas and related essential understandings provide the framework for our discussion of rational numbers. These are identified as a group below to give you a quick overview and for your convenience in referring back to them later. Read through them now, but do not think that you must absorb them fully at this point. The chapter will discuss each one in turn in detail.

Big Idea 1. Extending from whole numbers to rational numbers creates a more powerful and complicated number system.

> **Essential Understanding 1a.** Rational numbers are a natural extension of the way that we use numbers.

> **Essential Understanding 1b.** The rational numbers are a set of numbers that includes the whole numbers and integers as well as numbers that can be written as the quotient of two integers, $a \div b$, where b is not zero.

> **Essential Understanding 1c.** The rational numbers allow us to solve problems that are not possible to solve with just whole numbers or integers.

Big Idea 2. Rational numbers have multiple interpretations, and making sense of them depends on identifying the unit.

Essential Understanding 2a. The concept of *unit* is fundamental to the interpretation of rational numbers.

Essential Understanding 2b. One interpretation of a rational number is as a part-whole relationship.

Essential Understanding 2c. One interpretation of a rational number is as a measure.

Essential Understanding 2d. One interpretation of a rational number is as a quotient.

Essential Understanding 2e. One interpretation of a rational number is as a ratio.

Essential Understanding 2f. One interpretation of a rational number is as an operator.

Essential Understanding 2g. Whole number conceptions of *unit* become more complex when extended to rational numbers.

Big Idea 3. Any rational number can be represented in infinitely many equivalent symbolic forms.

Essential Understanding 3a. Any rational number can be expressed as a fraction in an infinite number of ways.

Essential Understanding 3b. Between any two rational numbers there are infinitely many rational numbers.

Essential Understanding 3c. A rational number can be expressed as a decimal.

Big Idea 4. Computation with rational numbers is an extension of computation with whole numbers but introduces some new ideas and processes.

Essential Understanding 4a. The interpretations of the operations on rational numbers are essentially the same as those on whole numbers, but some interpretations require adaptation, and the algorithms are different.

Essential Understanding 4b. Estimation and mental math are more complex with rational numbers than with whole numbers.

Accurate terminology is an important aspect of mathematics—and of this book. Suppose that someone asks you to use the word *nail* in a sentence. Would your sentence use the word as in, "I hit the nail on the head with my hammer," or as in, "I just broke my nail trying to open this jar," or neither? This example illustrates something that we are all familiar with—that is, many words have multiple meanings, but we can normally determine which meaning is intended by the specific context. It may come as a surprise, but many mathematical terms also have different meanings, depending on the context. This book deals with words such as *fraction, percent, rate,* and *unit,* all of which have multiple meanings in mathematics. It provides explanations, contexts, and examples, as well as formal definitions, to try to make the meanings clear.

Besides using the same word to mean different things, people sometimes use different words to mean the same thing. For example, consider the fact that some people use *rational number* and *fraction* interchangeably; however, these terms are not synonymous. We will define these two terms formally in our discussion of Essential Understanding 1*b*. Until then, however, we simply note that all rational numbers can be expressed in a fraction form, and so we begin our study of rational numbers by examining the history and uses of fractions.

Extending Our Use of Numbers: Big Idea 1

Big Idea 1. *Extending from whole numbers to rational numbers creates a more powerful and complicated number system.*

Have you ever wondered why we even have fractions? We sometimes hear arguments for eliminating their use. These pleas usually come from people who are frustrated by fractions, especially when they have to compute with them. Many have even advocated switching to the metric system simply so that we could, in their view, "get rid of" fractions. At some level, this frustration is understandable. It is true that concepts associated with fractions are more subtle and difficult than those needed to understand whole numbers. Nevertheless, pleas to abandon fractions are futile. Fractions are unavoidable; Reflect 1.1 offers a case in point.

Reflect 1.1

Six times a number is 42. What is the number?

Suppose that you change the first statement in the problem to "Six times a number is 32." Now what is the number?

Reflect 1.1 suggests one need for fractions. A very minor change in the problem eliminates the possibility of using whole numbers to solve it. In fact, only the counting numbers that are multiples of 6 give rise to a whole number answer; all others require additional rational numbers to solve the problem.

Fractions are here to stay. Perhaps the most compelling reason arises from the belief that algebraic skills are necessary for all students. The study of algebra is impossible without an understanding of fractions. Besides important ideas involving proportional reasoning that come up in algebraic contexts (see Lobato and Ellis [2010]), almost all instances of division in algebra are represented as fractions, and to work well with any of these expressions, we must have knowledge of fractions.

Even if your heart starts to palpitate when fractions are mentioned, you are no doubt comfortable with them at some level. Fractions pop up constantly in our daily lives: "It takes $3^1/_2$ hours to get there," "You will need $^2/_3$ cup of flour," "Today only, $^1/_2$ off," "I need a $^5/_8$-inch socket wrench to fix this," "Do you have these shoes in a $7^1/_2$; the 7 is too tight." We could go on with this list of everyday uses of fractions, but consider some common uses that are much less noticeable. For example, how did you get to work today?

Ideas related to proportional reasoning in algebraic contexts are discussed further in *Developing Essential Understanding of Ratios, Proportions, and Proportional Reasoning for Teaching Mathematics in Grades 6–8* (Lobato and Ellis 2010).

Did you drive a car, ride a bike, or take mass transit? Maybe you live close enough that you were able to walk? In any of these cases, you traveled at some rate of speed: maybe 65 mph on the freeway or 30 mph in town, 12 mph on your bike, or 2 mph walking. If you thought anything about how fast you were traveling, it is very likely that you were using fractions! For instance, the speed 65 mph is an abbreviation for 65 miles per hour, which is just another way of expressing the fraction $^{65 \text{ miles}}/_{1 \text{ hour}}$. We could continue to give more justifications for the impossibility of eliminating fractions, but it is more important to turn to the essential understandings related to the powerful, yet complex, rational number system characterized in Big Idea 1.

From counting to measurement

Essential Understanding 1a. *Rational numbers are a natural extension of the way that we use numbers.*

So how did we get here? Why were fractions invented? Certainly, they were around before algebra or computations of speeds using miles per hour. To gain a better understanding of fractions and rational numbers, it is useful to look at the historical development of those ideas. Natural (counting) numbers were created when societies needed to count. Counting objects was a natural activity in every known society. However, not every society had abstract symbols to represent these numbers. Evidence indicates that some societies counted with objects; to count sheep, for example, a shepherd might collect pebbles in a one-to-one correspondence with the sheep, and the pebbles became a record of the number of sheep. Other societies, like the Paiela of Papua New Guinea or the tribes of Paraguay in South America, did not have a need for large numbers, so they used "body counting"; that is, words that they used to represent body parts also represented numerical values. They would simply point to body parts to indicate numbers (Schmandt-Besserat 1999).

Most cultures' representations of numbers were not as extensive or complete as our current Hindu-Arabic system; many were not based on ten or even on place value. Many Native American tribes independently developed numeration systems that closely approximated base-twenty number systems (developed from counting both fingers and toes) and were largely finite because there was no need for numbers for very large quantities. The Winnebago Indians of the upper Midwest are said to have been able to use their numbers for quantities as large as 1,000,000. For quantities larger than that, they used metaphorical terms, such as "leaves on the trees," "stars of the heavens," "blades of grass on the prairie," or "sand on the lake shore" (Eels 1913). Before contact with the outside world, several Eskimo tribes had no designated number words for quantities beyond 20 (Murdoch 1890). In addition, many African tribes—most notably the Hottentots—had no numbers beyond 5 and used terms that talked about "many" for larger numbers. These examples show that numbers were invented because of a need in a civilization. If a culture was primarily a hunter-gatherer society with little trade and few cross-cultural interactions, large numbers were not needed. Furthermore, operations with numbers were also largely undeveloped (Crawfurd 1863).

Cultures that did interact and trade with others created number systems that were more sophisticated and robust. However, our modern Hindu-Arabic system of numeration did not instantaneously appear. Even the concept of zero did not develop for hundreds of years after the Greeks began using the (primarily) base-ten system. The invention of zero is credited to the Hindus of India (Bourbaki 1998). The Roman numeral system that we see in movie credits, among other contemporary contexts, never had a zero. In summary, we recognize that numeration systems varied significantly among different cultures and originated in a society's need to *count*.

The same numbers that societies used for counting, they also used to *measure*—to quantify distance, time, volume, and other attributes. Crude measurement can be accomplished by using whole numbers, but as cultures became more advanced, skilled, and interconnected, they needed greater precision in measurement. Over time, this led to the creation of number values that were no longer whole number values. Everyday modern experiences make this development quite easy to understand. For example, if you are replacing a broken water pipe, you will quickly discover that neither a value of 1 foot nor 2 feet suffices if you need the replacement to be $1^5/_8$ feet in length. Fractions were created to represent such numbers. However, this development did not occur overnight. Some societies, like the highly skilled Egyptians, wrote only fractions that are called *unit fractions*—that is, fractions with a numerator of 1. Other societies worked only with specific common fractions—not all fractions in general (Filep 2001).

The important idea is that the need for increased precision in measuring required non-counting numbers, so the emergence of fractions was inevitable. This change also extended the types of questions that numbers could answer—from not only "how many" ("How many sheep do we have?" "How many fish have you caught?" "How many times are you going to repeat the experiment?") to also "how much" ("How much water will that pail hold?" "How much carpeting do we need?" "How much time will it take to complete the task?").

For more on these ideas about the expansion from whole numbers to rational numbers, see *Developing Essential Understanding of Number and Numeration for Teaching Mathematics in Prekindergarten–Grade 2* (Dougherty et al. 2010).

The uses of fractions have evolved far beyond measurement. Other uses are discussed throughout this book, especially in connection with Big Idea 2.

**Big Idea 2**
Rational numbers have multiple interpretations, and making sense of them depends on identifying the unit.

Defining *rational number*

Essential Understanding 1*b*. *The rational numbers are a set of numbers that includes the whole numbers and integers as well as numbers that can be written as the quotient of two integers, a ÷ b, where b is not zero.*

Essential
Understanding 1*a*

Rational numbers are a natural extension of the way that we use numbers.

We need to provide more clarification for the terms that we will be using. We began our discussion of Essential Understanding 1*a* by talking about the *counting numbers,* sometimes called the *natural numbers.* Counting numbers are most easily defined by listing them: {1, 2, 3, ...}. The elementary school curriculum begins with this set of familiar numbers; as we have briefly seen, at least a subset of them has been a part of almost all known cultures. If we simply include the number 0 in this set, producing {0, 1, 2, 3, ...}, we then have the set of *whole numbers.* When we add the negatives of each of these numbers to the set, realizing that –0 = 0, the resulting set of numbers {... –3, –2, –1, 0, 1, 2, 3, ...} is the *integers,* and we have key benchmarks on a traditional *number line* (see fig. 1.1).

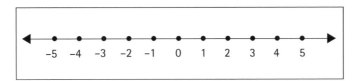

Fig. 1.1. A traditional number line with integers as benchmarks

Notice the space between adjacent integers on the number line in figure 1.1. Much of that space is taken by *rational numbers* that are not integers. A rational number is defined as a number that *can be* represented as the quotient, or indicated division, of two integers, as in *a* divided by *b,* with one important restriction: the divisor, *b,* must be different from zero. (We will discuss this restriction on *b* shortly.) This quotient can also be thought of as a ratio, and that link can help learners to remember the definition of the term <u>ratio</u>*nal number.* Symbolically, a rational number is often denoted by the form $a/_b$. So, for example, $3/_5$ is a rational number, as are $1/_3$, $2/_6$, $657/_{100}$, $^{-3}/_8$, and $6/_{-10}$. Reflect 1.2 poses a question about a quotient of two numbers, $a/_b$, when *a* or *b* or the quotient $a/_b$ itself is negative.

One important feature of the definition *of rational number* is that it includes any number that can be represented as a quotient of two integers. For example, the number $-{}^6/_{10}$ can be written as ${}^{-6}/_{10}$ or ${}^6/_{-10}$; therefore, $-{}^6/_{10}$ is a rational number; these are just three slightly different forms of the same number. This notion of *can be* means that every whole number is also a rational number; for example, 7 can be written as ${}^7/_1$ or as ${}^{21}/_3$. More broadly, any integer n can be written as ${}^n/_1$, so every integer is a rational number. Perhaps surprisingly, 1.45 is also a rational number; even though 1.45 is written as a decimal, it *can be* represented as ${}^{145}/_{100}$ (or ${}^{29}/_{20}$ or ${}^{435}/_{300}$). In other words, the fractions and decimals typically encountered in elementary and middle school represent rational numbers. The critical test to identify a rational number is not what it looks like but whether it *can be* written in the form of a quotient of two integers (with the divisor not equal to zero). Determining this becomes more difficult as numbers become more complicated: what about

$$\frac{5\frac{1}{4}}{47\frac{1}{2}}, \quad 2.333333, \quad \sqrt{2}, \quad \pi, \quad \text{and} \quad \sin 30°?$$

It is not immediately obvious which, if any, of these are rational numbers. Three of them are rational when we apply the stipulation that all rational numbers *can be* represented in a particular form. In Reflect 1.3, an everyday situation highlights a different requirement of the definition of *rational number*.

No zero denominators

The definition of *rational number* includes a condition that many people find puzzling: the denominator (the bottom number) of a rational number written in fraction form cannot be 0. Another way of saying this is that we cannot divide by 0. The natural question is, "Why can't we divide by 0?" Probably the most straightforward answer to this question begins with asking another question that most of us have no trouble answering: "What is 14 divided by 2?" All would agree that the answer is 7. When asked for a convincing argument why the answer is 7, many would respond, "Because 7 times 2 is 14." Formally, this argument is justified because multiplying by 2 is the *inverse operation* of dividing by 2. In general, this means that every division problem can be reformulated as a multiplication problem. Specifically, $^a/_b = \square$ can be rewritten as $\square \times b = a$. (Many textbooks describe this as the "missing factor" approach to division.) Now let's go back to the definition of a rational number and look at a specific example. Suppose that we did try to divide by 0. What would our answer be? Look at the statement $^8/_0 = \square$. If we rewrite this as a multiplication statement, we get $\square \times 0 = 8$. What could $\square$ equal? Since the product of any number and 0 is 0, there is no value for $\square$ that when multiplied by 0 would equal 8. To avoid this impossibility, we just say that we can never divide by 0, and consequently we can never have a rational number with 0 in the denominator. The careful reader might have noticed that although there is no real value that makes sense for $\square$ in the statement $\square \times 0 = 8$, there appears to be a value for $\square$ that would work in the statement $\square \times 0 = 0$, which is equivalent to $^0/_0 = \square$. In fact, any real number for $\square$ will work in the multiplication statement. This observation is why we say $^0/_0$ is *indeterminate*, but $^a/_0$ with *a* not equal to 0 is *undefined*.

Defining fraction

We stated previously that the word *fraction* has multiple meanings and is not synonymous with *rational number*. Now that we have defined *rational number*, it is time to define *fraction* formally in the way that we will henceforth use the word. *Fraction* is often used colloquially to mean "a little bit," as in the suggestion, "Move the picture down a fraction," or the response, "Just give me a fraction," to a question about how large a portion of dessert to serve. Until now we have allowed the word *fraction* to be interpreted loosely—even to be thought of as interchangeable with *rational number*. However, *fraction* really has a different meaning: a *fraction* is a *symbolic expression* of the form $^a/_b$ representing the quotient of two quantities (provided, as always, that the divisor *b* does not represent zero). Thus, all rational numbers expressed in the form $^a/_b$, like $^1/_3$, $^2/_6$, $^{657}/_{100}$, and $-^3/_8$, are immediately recognizable as fractions, but 1.45 is not. The number 1.45 *can be* expressed as a fraction ($^{145}/_{100}$), but as written, it is not a fraction. This interpretation of *fraction* (simply as an expression of the form $^a/_b$) means that it is an easy decision whether or not something is a fraction. You can simply look at it and decide.

By the very definition of *rational number*, all rational numbers *can be* written as fractions. However, there are also important fractions that are *not* necessarily rational numbers. For example,

$$\frac{\pi}{4}, \quad \frac{1}{x+3}, \quad \text{and} \quad \frac{\sin 45^\circ}{2}$$

are fractions, but they are not equivalent to the ratio of any two integers. (An exploration of the fact that $\pi/4$ can never be written in the form a/b, with a and b as integers, is beyond the scope of this book; a proof appears in Niven [1947].) With this interpretation of *fraction*, the algebraic expression $1/_{x+3}$ is clearly a fraction for all x not equal to –3, but depending on the value of the variable x, it may or may not be a rational number. Reflect 1.4 poses a question about how we regard different expressions of a whole number.

Reflect 1.4

Is 4 the same as 2 × 2 or 51 – 47 or *four*?

The question in Reflect 1.4 is confusing because of an ambiguity in language: what do we mean by "the same as"? Certainly, all four expressions represent the same value, but they are not exactly the same. The expressions 2×2 and 51 – 47 are both indicated operations that result in 4. The symbol "4" is a *numeral* that stands for the *abstract* number describing "fourness." In the same way, the fraction $1/_3$ is a numeral that represents a particular rational number.

With whole numbers, the number-numeral distinction can seem fussy and distracting, so it is often ignored. Similarly, the word *fraction* is often used to refer to the rational number that the fraction form represents. It is easy to see why we might be tempted to use it this way, since all rational numbers can be represented in fraction form. However, as we pointed out, not all fractions are rational numbers. For this reason, we restrict the term *fraction* to mean a symbolic expression of the form $a/_b$. One consequence of this definition is that $1/_2$ and $4/_8$ are different fractions, even though they both obviously represent the same number! The standard terminology is that they are *equivalent fractions,* meaning literally that they represent the *same value.*

Because the focus of this book is on rational numbers specifically rather than on fractions in general, we will place a further restriction on our use of the word *fraction*. In the rest of the book, unless a specific topic dictates otherwise, when we refer to *fraction*, we will mean *a symbolic expression of the form* $a/_b$ *representing a nonnegative rational number.* This definition includes less common forms, such as

$$\frac{1.2}{5.74} \quad \text{and} \quad \frac{2/3}{5/9}.$$

Most of our discussion will focus on simple forms like $^3/_8$, however. Justifications for our nonnegative restriction are that schoolchildren begin to study fractions long before they are introduced to the integers, that much of the arithmetic of rational numbers can be developed without reference to negative numbers, and that many of the ideas related to rational numbers are clearer in reference to positive fractions. In other words, with few exceptions, we will be using the word *fraction* in a nonstandard way, to refer to only a subset of the set of rational numbers, and likewise to a subset of all fractions more broadly defined (all quotients in the form $^a/_b$). However, aspects of fractions, and of rational numbers in general, discussed in this book extend easily and naturally to negative rational numbers.

One final point is that our use of the word *fraction* does not imply any one particular *interpretation* of a fraction, such as the part-whole interpretation. Big Idea 2 comprises many other interpretations.

Finding closure with rational numbers

Essential Understanding 1c. *The rational numbers allow us to solve problems that are not possible to solve with just whole numbers or integers.*

Extending our set of available numbers from integers to rational numbers also expands our ability to do arithmetic! You may have never thought about it in this way, but consider how a number system might be expanded. Using whole numbers, we can add or multiply any two whole numbers, and the result is another whole number. For example, the sum represented by 148 + 5793 and the product represented by 148 × 5793 are whole numbers. But this is not necessarily true if we want to subtract two whole numbers. Although the result of performing 9421 – 569 is a whole number, 569 – 9241 is not. In the primary grades, teachers sometimes say (inaccurately) that we can never subtract a larger number from a smaller number. The fact is that we certainly can, but to get an answer, we need to extend the set of numbers that we are working with from whole numbers to integers.

If we look at the last ordinary arithmetic operation that students formally encounter, division, we see that it is rarely the case that the result of dividing two whole numbers, or even two integers,

 Big Idea 2 *Rational numbers have multiple interpretations, and making sense of them depends on identifying the unit.*

gives another whole number or integer. The formal mathematical term for a guarantee that we will be able to stay within a given number system while performing an operation is *closure*. For example, the set of whole numbers has closure *under addition* but not *under subtraction,* as evidenced above in the subtraction 569 – 9241. Similarly, the set of integers is closed under addition, subtraction, and multiplication but is not closed under division; the result of 8 ÷ 3, for instance, is not an integer. To achieve closure under division, we must extend our number system again, this time to include the rational numbers. The operation suggested by 8 ÷ 3 has a result within this expanded system—namely, the number $^8/_3$. In fact, the division of any two rational numbers (disallowing zero divisors) results in another rational number. Similarly, the second question in Reflect 1.1 now has an answer: the result of 32 ÷ 6 is $^{32}/_6$. With this last extension, we now have a new and highly desirable property for our set of available numbers: the set of rational numbers is closed under all four operations—addition, subtraction, multiplication, and division (with the unavoidable exception that we still cannot divide by zero). Consider the usefulness of this extended set of numbers, with closure under the four operations, in solving problems like those in Reflect 1.5.

Reflect 1.5

1. Jane can run the 100-meter dash in 18 seconds. If she runs at a steady rate, how long would it take her to run 1 meter?

2. If gas costs $3 per gallon, how much gasoline can you purchase for $10?

3. What are the dimensions of a business envelope?

Problems like those in Reflect 1.5 exhibit how much better suited rational numbers are than either the whole numbers or the integers to many simple real-life applications. When we need to divide whole numbers in real-world contexts, we often get a rational number that is not a whole number. In problem 1, we need to take $^1/_{100}$ of 18 seconds, and we get $^{18}/_{100}$, or 0.18, seconds, which is less than 1 second. In problem 2, we need to divide 10 by 3, and we get $^{10}/_3$, or $3^1/_3$, gallons. Problem 3 is one example of the many practical measurement problems that require rational numbers.

Making Sense of Rational Numbers: Big Idea 2

Big Idea 2. Rational numbers have multiple interpretations, and making sense of them depends on identifying the unit.

In our discussion of Big Idea 1, we described what rational numbers are, why they are important, and how they came about. Turning now to Big Idea 2, we will show that rational numbers can be used in many different contexts, that their interpretation can change depending on the context, and that defining the unit is key to the interpretation (Behr et al. 1983; Carraher 1992, 1996; Kieren 1992; Lamon 2007). The following discussions of the essential understandings associated with Big Idea 2 introduce a variety of real-life examples and rational number models that are central to this big idea.

Big Idea 1
Extending from whole numbers to rational numbers creates a more powerful and complicated number system.

Unit as the basis for interpretation

Essential Understanding 2a. The concept of unit *is fundamental to the interpretation of rational numbers.*

Why is the unit so important? To describe the size of some quantity with a rational number, the first step is to determine what serves as the *unit,* or *whole.* Consequently, the rational number that results must be interpreted with respect to that unit. Why is this so?

Fractions may provide the clearest example, though the same principle applies to decimals. Figure 1.2 shows how the fraction describing the length of segment *M* differs, depending on what is designated as the unit. In each case, the line segment from 0 to 1 represents 1 unit of length. In example (a), the length of *M* is $1/_3$ because it is one of three equal-sized segments that make up the unit. In example (b), the length of *M* is $1/_2$ because two such segments constitute the designated unit.

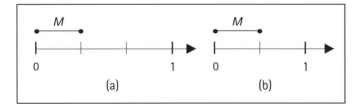

Fig. 1.2. Units and fractional parts

Some authors distinguish between *whole* and *unit.* Bassarear (2008) elaborates on this distinction. In informal contexts, including school, these terms are often used interchangeably, and we will not maintain a formal distinction in this book. We will, however, give preference to the term *unit* over *whole* to describe the quantity designated to be equal to 1 in work with rational numbers. We do this to try to avoid misconceptions that might arise from interpreting the word *whole* too narrowly. However, a disadvantage of the term *unit* is that it has multiple meanings; consider the term *unit fraction* (a fraction with a numerator of 1), for example.

Multiple interpretations: Rational numbers as part-whole relationships

Essential Understanding 2b. *One interpretation of a rational number is as a part-whole relationship.*

Your first introduction to rational numbers was probably through everyday experiences, such as putting half of your pennies into a piggy bank, folding a paper into thirds, or watching someone cut a pizza into eighths. When we talk about $^3/_8$ of a pizza, we are using fractions to describe a relationship between a part (represented as $^3/_8$) and a whole (represented as $^8/_8$). In the part-whole interpretation of rational numbers, a fraction indicates a comparison of the number of parts, indicated by the numerator, to a fixed number of equal-sized parts that compose the unit or whole, indicated by the denominator. So $^3/_8$ of a pizza typically indicates 3 pieces of a pizza that is cut into 8 equal-sized parts. Although the parts must be equal-sized, they do not have to be the same shape!

A unit can be a *continuous* (measurable) quantity, such as the area of a lawn, the length of a pencil, or the volume of water in a swimming pool, or a standard unit of measurement such as a square foot, a centimeter, and a gallon. The unit can also represent a group of *discrete* (countable) things, such a class of students, a set of pages in a book, or a set of tickets sold. In figure 1.3, the shaded portions in examples (a) and (b) can both be named as $^1/_3$ since each example shows 1 part out of 3 equal-sized parts. The darker portion in (a) illustrates $^1/_3$ of a unit used to measure a continuous attribute—the area of a region—and the darker portion in (b) illustrates $^1/_3$ of a unit composed of 12 discrete hearts. The rational number describing the darker part in each example represents a part-to-whole relationship. Reflect 1.6 explores the use of part-whole relationships in comparisons of quantities.

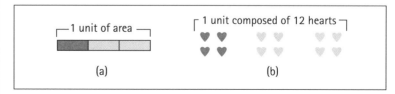

Fig. 1.3. One-third as one of three equal-sized parts of a unit

Reflect 1.6

In each example below, which of the two shaded areas represents more?

Example 1   Example 2

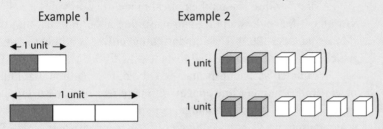

Is there more than one way to think about the meaning of this question? Explain.

The first question in Reflect 1.6 is ambiguous. One way to interpret the problem would result in directly comparing the sizes of shaded areas in example 1 or the numbers of shaded cubes in example 2. The more common interpretation, however, would lead to finding which of the shaded regions represents the greater fractional amount shaded. In this case, we would determine how much or how many are shaded *compared to the whole*. In both examples, half of the top unit is shaded and a third of the bottom unit is shaded. So the shaded portion in the top figure in each example represents the greater *fractional* amount. Understanding that the magnitude of a fractional number is relative to the size of the unit is a key to understanding rational numbers.

Multiple interpretations: Rational numbers as measures

Essential Understanding 2c. *One interpretation of a rational number is as a measure.*

When a rational number is interpreted as a measure, it describes the amount of something (like a distance, an area, a capacity or volume, or a duration in time) in relation to the size of a unit, which is considered equal to 1. If we say that a pitcher contains $1^3/_4$ cups of water, $1^3/_4$ is a measure of a quantity of liquid based on a unit measure of 1 cup.

In some measurement situations, the actual size of the unit is not specified but implied. For example, when we say that a runner has run $2/_3$ of a race, the implied unit is the length of the race. Even if we don't know how long the race is, we know how far the runner has already run compared to the total distance of the race. We could model the situation to show the relationship between these two distances without knowing whether the runner is running a hundred-yard dash or a marathon.

A number line is a mathematical model that provides a rich environment for understanding and reasoning about rational numbers (Moss and Case 1999). The standard number line is a linear scale, which is segmented into equal distances, with the distance from 0 to 1 designated as 1 unit. It may be curved or straight, horizontal or vertical. A number line appears in many real-life measuring devices, such as a ruler, rain gauge, cylindrical measuring cup, postal scale, thermometer, or speedometer (if it has a consistent scale). The unit may be partitioned into any number of subdivisions (fourths, twelfths, hundredths, millionths, and so on), according to the desired precision.

What makes a number line such a useful model?

A number line is a powerful tool for representing rational numbers, including positive and negative numbers. One reason is that it develops a sense of the relative magnitude of numbers by providing a visual image of how much greater $^5/_8$ is, say, than $^1/_2$, or how close 0.9 is to 1. Also, unlike materials such as pattern blocks or fraction bars, which work better with some particular forms of rational numbers and less well with others, the number line is flexible and can more easily be subdivided into parts of any size.

To identify a number denoting a point on a number line, we determine how far the point is from 0, relative to the unit interval (the distance from 0 to 1). Both the name of the point and the measure of the distance from 0 represent the same rational number, or *value*. So, for example, 1.6 is the name of a point that is 1.6 units to the right of 0 on a standard number line. To locate a point on a number line requires active thinking and reasoning, not just counting parts in a whole that has been already subdivided into a particular number of parts. The number line provides a context for understanding fractions that is different from, and more versatile than, part-whole models with the fractional parts already subdivided. This idea is explored in Reflect 1.7.

See Reflect 1.7
on p. 23.

To estimate the value represented by point *A* on the first number line in Reflect 1.7, we might first iterate the length of 1 unit along the number line. We would see that the value indicated by point *A* is greater than 2 units but less than 3 units. To determine a closer estimate of the value, we could partition the space between 2 and 3 on the number line into equal parts. We might be satisfied to say that the value represented by point *A* is about $2^1/_2$, or we could subdivide the unit into finer and finer equal subdivisions to

Reflect 1.7

Give a rough estimate of the values represented by points *A* and *B* on the number lines below.

How is using the number line model similar to, yet different from, a part-whole model?

get a more precise estimate. Naming rational numbers—estimating values—by using a number-line model with only a few labeled points, such as 0, 1, and some common benchmarks, reinforces an understanding of relative values of numbers and elicits complex reasoning.

A quantity compared to a whole

The interpretation of a rational number as a measure pushes us beyond our interpretation of a fraction as a *part of a whole* to the broader idea of a fraction as a *quantity compared with a whole*. This may seem like a small distinction, but it is an important one. For example, it makes sense to say that $^1/_4$ or $^3/_8$ is part of a whole. However, it may be confusing to say that $^5/_3$, or $1^2/_3$, is part of a whole. Instead, the fraction $^5/_3$ tells how much the *quantity* $^5/_3$, or $1^2/_3$, is *compared with* a whole. When comparing a given distance on a number line with the unit interval, it doesn't matter if the distance is greater than or less than 1 unit. A distance of 3.8 meters, for instance, is 3.8 times as much as 1 meter, which is defined as the unit.

Interpreting a fraction in terms of a unit fraction

Iterating unit fractions is another way to make sense of fractions and *improper fractions* (fractions with numerators that are larger than the denominators). Suppose that an arbitrary unit—any unit— is divided into 5 equal-sized pieces. Each piece represents the unit fraction $^1/_5$. Then the fraction $^3/_5$ can be thought of as "3 copies of the unit fraction $^1/_5$." This conception extends nicely to fractions larger than a whole, like $^{12}/_5$, which would be 12 copies of the unit fraction $^1/_5$. To generalize, the fraction $^N/_D$ could be interpreted as N copies of the fraction $^1/_D$.

Norton and McCloskey (2008) discuss the idea of unit fractions in more depth. In addition, these ideas are discussed with regard to whole numbers and measure in *Developing Essential Understanding of Multiplication and Division for Teaching Mathematics in Grades 3–5* (Otto et al., forthcoming) and *Developing Essential Understanding of Number and Numeration for Teaching Mathematics in Prekindergarten– Grade 2* (Dougherty et al. 2010).

Multiple interpretations: Rational numbers as quotients

Essential Understanding 2*d*. *One interpretation of a rational number is as a quotient.*

A rational number can also represent a *quotient* of two numbers. As a quotient, a rational number can be interpreted in two related ways. Take $^3/_4$, for example. First, it can indicate a division operation: that is, $^3/_4$ can be seen as equivalent to the arithmetic expression $3 \div 4$. Second, it can mean the single number that results from performing this operation, one form of which is $^3/_4$, or in decimal form, 0.75.

Consider an example that shows what $^3/_4$, interpreted as 3 divided by 4, might represent in an everyday scenario. Suppose four speakers are giving a presentation that is three hours long; how much time will each person have to present if they share the presentation time equally? One way to solve this problem is to divide each hour into fourths and then divvy up the fourths so that each speaker has an equal amount of time (see fig. 1.4). Each person would get three $^1/_4$ hours, or $^3/_4$ of an hour. Whenever we divide something 4 ways, one share will be equal to $^1/_4$ of the original amount. Notice that each speaker's share ($^3/_4$) is equal to $^1/_4$ of 3 hours. Reflect 1.8 explores other number-line approaches to the problem.

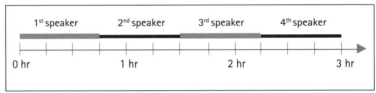

Fig. 1.4. A fraction as a quotient

Reflect 1.8

Continue to consider the scenario of four speakers sharing three hours of presentation time.

Show another way (besides that shown in fig. 1.4) to use a number line to divide 3 units into 4 equal parts.

Talk with your colleagues about how to find the amount of 1 share. What is the amount of 1 share?

The units representing presentation hours on the number line may be divided in many ways. For example, the first two hours could be divided into halves, giving each of the four people $^1/_2$

hour. Then the last hour could be divided into fourths, giving each person $1/4$ of an hour more. As long as the total presentation time is divided into equal shares, each person will get $3/4$ of an hour.

Interpreting a rational number as a quotient doesn't make sense unless we agree that we can divide a "smaller number" by a "larger number." Many of us were told differently when we were younger. However, it makes sense that if the number of presentation hours (3) is fewer than the number of people sharing the time (4), then each share will be less than 1 hour. In other words, the quotient $3 \div 4$ will be a number $(3/4)$ that is less than 1.

Multiple interpretations: Rational numbers as ratios

Essential Understanding 2e. *One interpretation of a rational number is as a ratio.*

A *ratio* expresses a relationship between two (and sometimes more) quantities or parts of quantities and compares their relative measures or counts. The ratio $2/15$, for example, could represent a ratio of 2 adult chaperones for every 15 teenagers, and $3/5$ might refer to 3 cups of rolled wheat for every 5 cups of rolled oats in a recipe for granola. Note that the order of the numbers in a ratio is critical. In the last example, the ratio comparing rolled wheat to rolled oats is $3/5$, whereas the ratio comparing rolled oats to rolled wheat is $5/3$. Like other interpretations for rational numbers, one part of the ratio is always related to the other part by multiplication. In the granola ratio, the volume of rolled wheat is $3/5$ times the volume of rolled oats, regardless of the actual quantities (e.g., 6 cups to 10 cups) or the units of measure involved (cups, liters, tablespoons, or something else). The comparison is multiplicative—not additive.

Expressing a ratio as a percent

Another way to express a ratio is as a *percent*. One meaning of *percent* is a "per 100" relationship; $N\%$ can be interpreted as a ratio of the form $N/100$, where N can be a whole number, a fraction, a decimal, or a mixed number. For instance, 3.2% can indicate a ratio where the numerator is the given number (3.2) and the denominator, indicated by the percent sign (%), is 100; thus, 3.2% can be written as $3.2/100$. If the ratio of students accepted into a university is 2 for every 5 students who apply, then the ratio is equivalent to $40/100$, so the acceptance rate is 40% (40 accepted students per 100 applications).

When two ratios can be meaningfully compared, percents simplify the numerical comparison by standardizing the denominator.

For example, it may not be obvious that 7 successful shots out of 20 free throws is a better record than 8 out of 25. However, when these ratios are expressed as percentages, 35% (7 out of 20, which is equivalent to 35 out of 100) versus 32% (8 out of 25, equivalent to 32 out of 100), it is easy to determine which is the better record. Chapter 2 provides additional discussion of percents.

Rates, ratios, and percents

Developing an Essential Understanding of Ratios, Proportions, and Proportional Reasoning for Teaching Mathematics in Grades 6–8 (Lobato and Ellis 2010) discusses ideas about *rate* in greater detail.

One interpretation of *rate* is as a set of equivalent ratios maintaining a given multiplicative relationship (Thompson 1994). Rates often compare quantities of two different kinds of units. For example, a rate of speed such as 30 miles per hour may be expressed as 30 miles per 1 hour, $^{30 \text{ miles}}/_{1 \text{ hour}}$, or more simply, $30 \, ^{\text{miles}}/_{\text{hour}}$. This rate can also be expressed by an infinite number of equivalent ratios including

$$\frac{60 \text{ miles}}{2 \text{ hours}}, \quad \frac{15 \text{ miles}}{0.5 \text{ hours}}, \quad 0.5 \, \frac{\text{miles}}{\text{minute}}, \quad \text{or } 720 \, \frac{\text{miles}}{\text{day}}.$$

Other examples of rates are $^{\$15}/_{\text{cubic yard}}$, 3 chapters every 4 weeks, $3.25 per pound, and 1.3 minutes per lap. In school mathematics, a rate can also compare two different measures that use the same unit, such as an interest rate of $4 charged for every $100 borrowed, usually expressed as 4%. Rates are used primarily to describe situations where two quantities may change in magnitude but always maintain a constant ratio (also see Lobato and Ellis [2010]).

How are fractions and ratios related?

The concept of ratio is explored in depth in *Developing an Essential Understanding of Ratios, Proportions, and Proportional Reasoning for Teaching Mathematics in Grades 6–8* (Lobato and Ellis 2010).

A fraction $^a/_b$ is commonly used to represent a ratio that is a part-to-whole comparison. However, it is also used to represent a ratio that is a part-to-part comparison, as in the granola recipe above, where we had

$$\frac{3 \text{ cups rolled wheat}}{5 \text{ cups rolled oats}},$$

or to represent a quantity-to-quantity comparison, as in the price of bananas, for example,

$$\frac{\$3}{5 \text{ pounds}}.$$

Part-part ratios are easy to turn into part-whole ratios. If we assume that the only grains that our granola contains are rolled wheat and rolled oats, then the total amount of grain is 8 cups (or liters or tablespoons), of which 3 are rolled wheat and 5 are rolled oats. Then the part-whole ratios describing the portions of rolled wheat and rolled oats in the granola can be expressed as the fractions $^3/_8$ and $^5/_8$, respectively (see Lobato and Ellis [2010]).

Multiple interpretations: Rational numbers as operators

Essential Understanding 2f. *One interpretation of a rational number is as an operator.*

A rational number can also be used as an *operator,* which changes or transforms another number or quantity by magnifying, shrinking, enlarging, reducing, expanding, or contracting it. With rational numbers, the number itself is the operator, and the implied operation is multiplication. Both of the following situations describe a change of size or scale that results when the number is used as an operator:

- The amount of waste is $^3/_4$ as much this year as last.

- Cereal A has 2.5 times as much sugar as cereal B.

For a visual example, consider figure 1.5, which shows $^2/_3$ acting as an operator on the length of strip A, resulting in a length that is $^2/_3$ as long, illustrated by strip B. Reflect 1.9 investigates the relationship between strips A and B further.

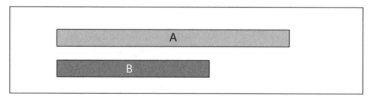

Fig. 1.5. The effect of the operator $^2/_3$ on a length A

Reflect 1.9

Referring to figure 1.5, describe the length of strip A compared to that of strip B, assuming that $^2/_3$ is an operator acting on strip A.

How long is strip B compared to strip A?

What are some possible lengths for strips A and B?

To gain further insight into these questions, let's examine the algebraic equation $B = (^2/_3)A$. A fraction acting as an operator combines the opposing influences of its numerator, which has the effect of a multiplier, and its denominator, which has the effect of a divisor. For a simple fraction $^a/_b$ where a and b are whole numbers greater than 1, the numerator essentially expands the number or quantity being operated on, while the denominator shrinks it. The net effect of these two operations acting together in a rational number operator depends on the size of the numerator compared with the size of the denominator. If both are the same number, as in $^3/_3$, for instance, the numerator essentially multiplies by 3 while the denominator divides by 3, so the size of the number or quantity being operated on by $^3/_3$ (or 1) remains unchanged. If the value of the fraction operator is greater than 1, the effect of the larger numerator "outweighs" the smaller denominator, and the operation magnifies the original quantity; similarly, a fraction less than 1 scales it down.

The net result of a simple fraction acting as an operator will always be the same whether we enlarge first (by multiplying by the numerator) and then shrink (by dividing by the denominator), or shrink first and then enlarge, or apply both at once by simply multiplying by the fraction. For example, if we have a length that is 12 inches and are using $^2/_3$ as an operator, we can find the result by first doubling 12 (to get 24), and then dividing by 3 (to get 8); or by first dividing 12 by 3 (to get 4), and then multiplying by 2 (to get 8). Or we can multiply 12 by the fraction $^2/_3$ to get 8.

An expanded view of *unit*

Essential Understanding 2g. *Whole number conceptions of* unit *become more complex when extended to rational numbers.*

When we first learn about fractions, we understand that a unit can be a single thing, such as a submarine sandwich (see fig. 1.6a). We also need to understand that a unit can be a collection of discrete things that are *unitized,* or mentally "chunked" together as a unit (Steffe 2001). A litter of puppies is an example of a unit consisting of discrete objects that are mentally chunked together (see fig. 1.6b). The context of the situation determines the unit. So when someone says that $^2/_5$ of a litter of puppies is female, we know from the context of the statement that the number of puppies in the litter is the unit (see also Dougherty and colleagues [2010]).

For a discussion of unit as related to whole numbers, see *Developing an Essential Understanding of Number and Numeration for Teaching Mathematics in Prekindergarten– Grade 2* (Dougherty et al. 2010).

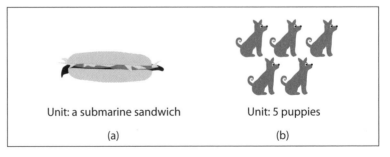

Unit: a submarine sandwich   Unit: 5 puppies

(a)   (b)

Fig. 1.6. Examples of a unit (a) as a single thing and (b) as a group of things

An object that is composed of a number of pieces is sometimes called a *composite unit*. Examples of composite units include a carton of eggs and a sheet of stamps (see fig. 1.7).

Fig. 1.7. A carton of eggs and a sheet of stamps as composite units

A unit can also be a measure, which might be represented by a whole number, a fraction, or a mixed number. For example, suppose a plant is now $1\frac{1}{2}$ feet tall, but it was only $\frac{1}{4}$ as high when you planted it. The unit can be conceptualized as $1\frac{1}{2}$ feet, which is the current height of the plant (see fig. 1.8). So $\frac{1}{4}$ of the unit tells how high the plant was when it was planted. On the other hand, the plant's initial height, $\frac{3}{8}$ feet, could be designated as the unit. In this case, a natural question might be to ask how much the plant has grown with respect to its beginning height.

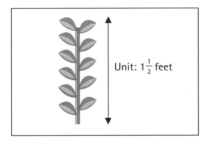

Unit: $1\frac{1}{2}$ feet

Fig. 1.8. A unit as a measure

Most of the time a unit is explicit in the context. Sometimes, however, the unit is implicit and must be inferred. For example, if we know that a third of a casserole is left, we can infer the total amount of casserole (the unit) before any was eaten. The whole casserole, or unit, is 3 times as much as $\frac{1}{3}$ (the amount left). Lamon (1999) provides a detailed discussion about different forms of units.

This discussion of the essential understandings related to Big Idea 2 has touched on several ideas that the discussion of Big Idea 3 and its associated understandings will elaborate further. As you read more about fractions and decimals, keep in mind that the unit in its various forms is a key idea in understanding the diverse interpretations of rational numbers.

 **Big Idea 2**
Rational numbers have multiple interpretations, and making sense of them depends on identifying the unit.

 **Big Idea 3**
Any rational number can be represented in infinitely many equivalent symbolic forms.

Rational Numbers and Equivalence: Big Idea 3

Big Idea 3. *Any rational number can be represented in infinitely many equivalent symbolic forms.*

A core concept in mathematics is equivalence. Two numerical expressions that represent the same quantity are *equal* to one another and are called *equivalent* expressions.

Research shows that many students who clearly state that they understand that 2 + 3 = 5 do not believe that 5 = 2 + 3 or even that 5 = 5 (Falkner, Levi, and Carpenter 1999). When young students work with whole numbers, the first written examples of equivalence that they see usually include an expression indicating an operation on the left-hand side, such as 5 + 3 = 8 or 5 × 3 = 15. As a result, students tend to interpret the equals sign as a signal to do something—namely, to perform the operation on the left to achieve the result on the right. However, this interpretation is not helpful for making sense of statements of equality like the following:

$$8 = 5 + 3 \quad 5 + 3 = 3 + 5 \quad 5 + 3 = 4 + 4 \quad 1.5 = \frac{15}{10} \quad \frac{4}{6} = \frac{8}{12}$$

These statements call for a more mathematical interpretation of the symbol for equality—the equals sign. It tells us that the numerical expression on the left-hand side represents the same quantity as that on the right-hand side. Because equivalence and equality are so important in understanding and working with rational numbers, we should not shortchange these concepts by spending too little time on them.

Representing rational numbers as equivalent fractions

Essential Understanding 3a. Any rational number can be expressed as a fraction in an infinite number of ways.

As we have seen, a fraction $^a/_b$ (where a and b are both nonnegative integers and b is not equal to zero) represents a single rational number. Two different fractions, though, can represent the same rational number.

Multiplying the numerator and denominator by the same number

Suppose that we cut a board into two parts of equal length by cutting it at the "halfway point," and we name each part $^1/_2$ to designate one of two pieces. If we then subdivide each of those halves into three parts of equal length, we will have divided the original board into six pieces of equal length. Each *half* of the original board will now contain 3 of the 6 pieces. The upper diagram in figure 1.9 shows a segment of a number line with the midpoint representing the rational number $^1/_2$, and the lower diagram shows the same segment divided into six equal parts. The point that separates the two halves, $^1/_2$, is precisely the same as the point named $^3/_6$. These points are exactly the same distance from 0.

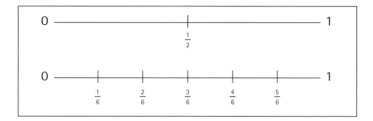

Fig. 1.9. Equivalent fractions $^1/_2$ and $^3/_6$ on a number line

In our hypothetical situation with the board, we first cut two equal lengths, as illustrated in the upper diagram in the figure. We designated each of these pieces $^1/_2$, and then we tripled the number of equal pieces (expressed in the denominator) from 2 to 6 and tripled the number of pieces under consideration (expressed in the numerator) from 1 to 3, as illustrated in the lower diagram in the figure. That is, we multiplied both the numerator and denominator by 3. This process of creating a new fraction that is equivalent to $^1/_2$ is illustrated by the following equation:

$$\frac{1 \times 3}{2 \times 3} = \frac{3}{6}$$

The terminology can be confusing. The two expressions $^1/_2$ and $^3/_6$ are different fractions, or more precisely, different fraction names for the same rational number. We usually read the statement $^1/_2 = ^3/_6$ as "$^1/_2$ equals $^3/_6$," but we commonly refer to $^1/_2$ and $^3/_6$ as "equivalent fractions." In the world of fractions, and of rational numbers in general, the terms *equal* and *equivalent* mean essentially the same thing (though in other contexts they may have distinct meanings).

Let's look at equivalence in a more general way, using what we know about multiplication. The rational number that can be

Note that the process described for creating equivalent fractions is not the same as the multiplication of rational numbers, which will be developed in the discussion of Big Idea 4.

 Big Idea 4
Computation with rational numbers is an extension of computation with whole numbers but introduces some new ideas and processes.

represented by $^1/_2$ has many fraction names; $^3/_6$ is just one of them. In fact, $^1/_2$, like every other fraction, has an infinite number of equivalent fractions. Imagine that we subdivided each of the two original equal parts of the board into *N* pieces of equal length. What would happen if we multiplied both the numerator and the denominator of the original fraction, $^1/_2$, by *N* (that is, by any counting number greater than 1)? We would obtain a new fraction that is equivalent to the original. This reasoning suggests a way to create lots of equivalent fractions.

Multiplying a number by 1

What happens when we multiply any number by the special number 1? First, let's recall that to multiply two fractions, we multiply the numerators together and the denominators together—for instance, $^3/_4 \times ^5/_7 = ^{15}/_{28}$. (The rationale for this procedure appears in the discussion of Essential Understanding 4a.) In the previous example of the board, we multiplied both the numerator and denominator of $^1/_2$ by 3, which is the same as multiplying $^1/_2$ by $^3/_3$. But $^3/_3$ equals 1! In other words, when we multiply both the numerator and denominator of a fraction by the same nonzero number, we are actually multiplying the whole fraction by the number 1:

$$\frac{1 \times 3}{2 \times 3} = \frac{1}{2} \times \frac{3}{3} = \frac{1}{2} \times 1$$

Essential
Understanding 2*f*

*One interpretation of a
rational number is
as an operator.*

Essential Understanding 2*f*, which states that we can interpret a rational number as an operator, can help us see that when we multiply any number by 1—the *multiplicative identity*—we get the same number back, even if it happens to be in a different form. When we use a rational number as an operator, the implied operation is multiplication, the numerator acts as an "expander" (multiplier), and the denominator acts as a "shrinker" (divisor). In the example above with $^1/_2$, if we interpret $^3/_3$ (which we recognize as another name for 1) as an operator, the numerator expands ($^1/_2 \times 3 = ^3/_2$) and the denominator shrinks ($^3/_2 \div 3 = ^3/_6$), with the result giving back our original number, $^1/_2$, though in a different form.

As the example of cutting the board demonstrated in the case of the fractions $^1/_2$ and $^3/_6$, when we multiply both the numerator and denominator of a fraction $^a/_b$ by the same number, the result is equivalent to $^a/_b$, although it may have a different fraction name. Reflect 1.10 explores this process with addition instead of multiplication.

Reflect 1.10

What happens if, instead of multiplying, you *add* the same number to both the numerator and denominator of a fraction?

Does $\dfrac{1}{2}$ equal $\dfrac{1+3}{2+3}$?

What happens if you choose other numbers?

In the example in Reflect 1.10, the result, $^4/_5$, is clearly not equal to $^1/_2$. Only in special cases will the original fraction and the resulting fraction be equivalent when the same number is added to the numerator and the denominator. In general, only by multiplying the numerator and denominator by the same nonzero number (i.e., multiplying the entire fraction by 1) can we generate an equivalent fraction.

Which is the "best" form of a fraction?

Although infinitely many fractions all represent the same rational number, one of those fractions may be more suitable than the others for specific purposes or contexts. For example, to find an equivalent percent or decimal, we may want our fraction to have a denominator of 10, or some power of 10. To add, subtract, or compare two fractions, we may want both fractions to have the same denominator. For instance, a carpenter who needs to know the combined thickness of two boards, one with a thickness of $2^5/_8$ inches and the other with a thickness of $1^3/_4$ inches, may find that the form $1^6/_8$, or even $1^{12}/_{16}$, depending on the folding carpenter's rule at hand, is more useful than $1^3/_4$.

   We sometimes want a fraction to be expressed in *simplest form*—that is, with both the numerator and denominator as counting numbers that have no common factors other than 1. Given a fraction whose numerator and denominator are counting numbers, we can find the equivalent fraction in simplest form by dividing both the numerator and the denominator by the same whole number factor (greater than 1). If necessary, we can repeat this process with additional factors, until no common factor greater than 1 remains. For $^5/_{10}$, for example, we would divide both the numerator and denominator by 5 (a *common factor*) to get $^1/_2$. For $^{210}/_{360}$, we could first divide both numerator and denominator by 10 to get $^{21}/_{36}$, then by 3 to get $^7/_{12}$. Or we could first divide by 3 to obtain $^{70}/_{120}$, then by 2 to get $^{35}/_{60}$, and finally by 5 to yield $^7/_{12}$, the equivalent fraction in simplest form.

Comparing fractions

As we have seen, we can use various methods to learn whether or not two fractions are equivalent. If two fractions do not represent the same rational number but are fractions of the same unit, then one must be greater than the other. We have several ways to determine which is greater. For example, suppose we want to know which of the two fractions $^5/_6$ and $^5/_8$ is greater. Probably the most popular method is to change both fractions to equivalent forms with a common denominator. We can conclude that $^5/_6$ is greater than $^5/_8$ because $^{20}/_{24}$ is greater than $^{15}/_{24}$. Alternatively, since these fractions have a common numerator, we have another way to reason about which is greater. First, we might ask which fractional piece is larger—sixths or eighths? If the unit is the same for each fraction, then dividing that unit into 8 equal parts will result in smaller pieces than dividing it into 6 equal parts (see fig. 1.10). Then, because the numerators tell us that we have 5 pieces of each (sixths and eighths), we can conclude that $^5/_6$ is greater than $^5/_8$.

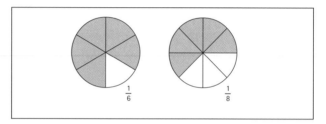

Fig. 1.10. Comparing unit fractions with different denominators

A third method involves comparing both fractions to benchmarks. We might think of $^5/_6$ as being closer to 1 and $^5/_8$ as closer to $^1/_2$ by reasoning that $^5/_6$ is $^1/_6$ less than 1 but $^2/_6$ greater than $^1/_2$, and $^5/_8$ is $^1/_8$ greater than $^1/_2$ but $^3/_8$ less than 1.

Rational numbers and density

Essential Understanding 3b. *Between any two rational numbers there are infinitely many rational numbers.*

We have seen that any rational number has many fractional forms, called *equivalent fractions*. Also, we have noted that if two fractions are not equivalent, one must be greater than the other. These important ideas bring us to the next topic, *density*. Reflect 1.11 introduces this topic.

Reflect 1.11

Find three different rational numbers between $\frac{11}{13}$ and $\frac{12}{13}$.

The rational numbers help to fill the gaps between the whole numbers. As we have seen, rational numbers allow us to describe measures that do not correspond exactly to whole numbers. So between 2 and 3, for example, we easily find $2\frac{1}{2}$, $2\frac{5}{11}$, and $2\frac{6}{11}$ (or, alternatively, $\frac{5}{2}$, $\frac{27}{11}$, and $\frac{28}{11}$). But a related idea may be surprising.

To approach this idea, let's begin with a question: Can we find a rational number between $\frac{27}{11}$ and $\frac{28}{11}$? See figure 1.11. On the one hand, it may seem as though no named rational number can work, but on the other hand, it looks as though there is room for other such numbers. Actually, there is lots of room—so much that it's hard to imagine!

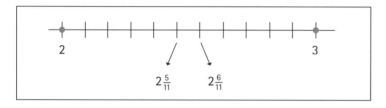

Fig. 1.11. Space between two rational numbers on the number line

A natural first response—and one that is often discounted because it appears to violate "proper form"—is,

$$\text{How about } \frac{27\frac{1}{2}}{11} \; ?$$

Is the rather odd numerical expression in this question a rational number? To decide, we need to determine whether the expression *can be* written as the quotient of two integers. We can simply double both the numerator and the denominator of this *complex fraction* (an expression of the form $\frac{x}{y}$ where at least one of x and y is itself a fraction and y is not zero) to obtain $\frac{55}{22}$, and we have found a rational number exactly halfway between $\frac{27}{11}$ and $\frac{28}{11}$!

Now we come to the surprising part. We can take this idea of multiplying both the numerator and denominator by the same factor and repeat the process to find that our unusual first expression is also equivalent to $\frac{110}{44}$ and $\frac{1100}{440}$ and $\frac{11000}{4400}$, and so on. Since $\frac{27}{11} = \frac{10800}{4400}$ and $\frac{28}{11} = \frac{11200}{4400}$, we see that we can find lots of rational numbers between $\frac{10800}{4400}$ and $\frac{11200}{4400}$, and all of them will be between $\frac{27}{11}$ and $\frac{28}{11}$! In fact, we could continue the process and find as many rational numbers between the original two rational numbers as we want.

Mathematicians use the word *density* to name this property of rational numbers—that between any two rational numbers we can find a third rational number. This property leads to the fact that between any two rational numbers, no matter how close, there are infinitely many other rational numbers. From a measurement standpoint, this means that rational numbers allow us to gain as much precision as we desire.

If we imagine all the rational numbers graphed as points on a number line, we will see that the number line no longer contains only the discrete (separated) points corresponding to the integers, with gaps between every two adjacent integers. Instead, each of these "gaps" contains an infinite number of points, with each point corresponding to a distinct rational number. Although this number line might now seem to be solid, or *continuous,* it still contains "holes" that the real number line fills with *irrational numbers,* such as π and the square root of 2. Chapter 2 provides a more detailed discussion and explanation.

Representing rational numbers in decimal form

Essential Understanding 3c. *A rational number can be expressed as a decimal.*

So far, we have discussed fractions in some detail. Rational numbers are often expressed in fractional form, but we also express them in decimal form. In fact, in our global society, decimals are the most common form in which we encounter rational numbers. Decimals are essential to metric measurement and are extremely useful in science and technology.

Children's earliest encounters with decimal points are likely to be in the context of money; young children probably first think of an expression such as $2.35 as representing two different *units*—that is, 2 *dollars* and 35 *cents*. Interpreted as a decimal, however, $2.35 has just one unit: the dollar. The decimal has a whole number part to the left of the decimal point, and, to its right, a part representing less than one unit. So the quantity $2.35 could be read as "two and thirty-five hundredths dollars." Using fraction forms, we have 2 dollars and $^{35}/_{100}$ of a dollar, or $2\,^{35}/_{100}$ dollars.

History

You may be surprised to learn that the familiar Hindu-Arabic numeration system that we use for whole numbers was not extended to include decimals for more than 1400 years. The Arabs and the

Chinese had been using decimals for centuries when, in 1585, Simon Stevin became the first European to publish conceptual details of decimals and to appreciate their significance. In 1670, Gabriel Mouton, a French theologian and mathematician, proposed the general use of the decimal system and suggested a standard linear measurement divided decimally—that is, by 10.

In 1790 Thomas Jefferson proposed a decimal-based measurement system to the United States House of Representatives. Congress gave this plan serious consideration, but ultimately it lost by one vote. The American founders, however, went ahead and adopted the first completely decimal currency system in 1792. Jefferson forwarded the idea that the hundredth part of the dollar be called a *cent,* after the Latin word *centum* for "one hundred," and that the tenth of the dollar be called a *dime,* from *decima* (or *decima pars*), meaning *tenth* or *tenth part* in Latin.

This and other information about the history of decimals and the decimal system is available at http://didyouknow .org/decimal/ and http://www.france -property-and -information.com /metric-system-and -history.htm.

These and other details about Jefferson's interest in a decimal-based currency can be found at http://wiki .monticello.org /mediawiki/index.php /Currency.

Properties

Our numeration system for whole numbers is built on four basic properties: a base-ten property, a positional property, and both multiplication and addition properties (Ross 1989). Let's take a look at those four properties in relation to a sample whole number—say, 527—with the help of the chart in figure 1.12. The single *digits* in the top row represent the "face values" corresponding to each position in the number. The bottom row tells the positional value of each digit— its *base-ten place value.* We *multiply* the face value of each digit by its place value, and then we *add* the results to get the number:

$$(5 \times 100) + (2 \times 10) + (7 \times 1) = 527$$

Face value	5	2	7
Place value	Hundred	Ten	One

Fig. 1.12. A chart using 527 to illustrate place value
with whole numbers

Decimals use the same basic numeration system as whole numbers, with 1 as a unit. The digits to the right of the decimal point have positional values that continue the base-ten place value pattern: as we move to the right, the value of each position is $1/10$ of that of the previous position. The chart in figure 1.13 and the expanded notation form below the chart summarize how this system works for the number 527.384. First, we multiply each digit by its positional value (5×100, 2×10, and so on, to $4 \times 1/1000$).

Then we add together all the products (500 + 20 + ... + $^4/_{1000}$) to get 527.384. The part to the right of the decimal point is read as "three hundred eighty-four thousandths" (or simply as a string of digits, "point three eight four"). Reflect 1.12 poses an obvious question, leading to our next topic.

Face value	5	2	7	.	3	8	4
Place value	100	10	1		$\frac{1}{10}$	$\frac{1}{100}$	$\frac{1}{1000}$

$$527.384 = (5 \times 100) + (2 \times 10) + (7 \times 1) + (3 \times \tfrac{1}{10}) + (8 \times \tfrac{1}{100}) + (4 \times \tfrac{1}{1000})$$

Fig. 1.13. A chart using 527.384 to illustrate place value with decimals

Reflect 1.12

Can *all* rational numbers be expressed as decimals?

What forms do the decimals take?

Fractions to decimals

When the numerator of a fraction is a whole number and the denominator is a power of 10, as in the fraction $^{180}/_{1000}$, finding the fraction's decimal equivalent is relatively straightforward. Thousandths require three decimal places, so the decimal form is 0.180. When a number is between 0 and 1, we use a leading zero in the decimal form for clarity. For fractions with denominators other than powers of 10, we might be able to find an equivalent fraction with a denominator that *is* a power of 10 and with an integer numerator, so that we can easily write the fraction as a decimal. Some are simple. For example, $^1/_2$ is equivalent to the fraction $^5/_{10}$, which equals 0.5.

How would you find the decimal equivalent for $^5/_8$? One interpretation for $^5/_8$ is the quotient that results from dividing 5 by 8. Whether you use a traditional paper-and-pencil algorithm or a calculator to perform the indicated division, you can see the resulting number, 0.625 (see fig. 1.14). This is called a *terminating decimal—* that is, a decimal that terminates after some fixed number of digits. (You could, however, append zeroes at the end to get 0.6250, or 0.62500, and so on, and this possibility has implications for precision in measurement contexts. Chapter 2 discusses precision in more detail.)

```
        0.625
    8) 5.000
        48
        20
        16
        40
        40
         0
```

Fig. 1.14. The equivalent decimal for $\frac{5}{8}$

But what is the decimal equivalent of $\frac{1}{3}$? When you divide 1 by 3, you get a *repeating decimal,* 0.3333333.... (The ellipsis indicates that the decimal continues and is often read simply as "repeating.") A repeating decimal is one that, from some position onward (when read from left to right), consists of a fixed finite sequence of digits that repeats an infinite number of times. The repeating sequence, called the *repetend,* may consist of just one digit or any finite number of digits. For instance, $\frac{7}{22}$ is 0.3 18 18 18...; in decimal form, the repetend is the digit sequence "18." A symbol called a *repetend bar* is sometimes used to show the finite sequence of digits that repeats, as in $0.3\overline{18}$. Patterns develop in repeating decimals, leading to important ideas. Reflect 1.13 investigates one of these.

Reflect 1.13

To convert the fraction $\frac{3}{7}$ into decimal form, we can divide 3 by 7.

What pattern develops when we use the long-division algorithm to do this? Why?

When we use pencil and paper to perform this computation, we see the repeating decimal begin to develop as shown in figure 1.15. The result of dividing 3 by 7 is 0.428571 428571 428571... The repetend has six digits, one less than the divisor. Is this a coincidence? When we divide one counting number by another, as in the rational number $\frac{p}{q}$, we can get only q possible remainders (from 0 up to $q - 1$). If q is 7, for example, the possible remainders are 0, 1, 2, 3, 4, 5, or 6. Each step in the long division results in one of these remainders. At some point, if the remainder is ever zero, the decimal terminates. If not, then eventually it must begin repeating remainders in a fixed pattern, and so the digits in the decimal quotient also begin a repeating pattern. Because there are only $q - 1$ *nonzero* remainders, the recurring pattern can contain no more than $q - 1$ digits.

```
                              Step by Step
        .4285714
     7)3.0000000              30/7 = 4 r 2  ←┐
       28                                    │
        20                    20/7 = 2 r 6   │
        14                                   │
         60                   60/7 = 8 r 4   │
         56                                  │
          40                  40/7 = 5 r 5   │
          35                                 │
           50                 50/7 = 7 r 1   │
           49                                │
            10                10/7 = 1 r 3    │
            07                                │
             30               30/7 = 4 r 2 ─┘  (again)
             28
              20
             etc.
```

Fig. 1.15. Steps in dividing 3 by 7

The important idea that this analysis illustrates is that *any* rational number can be expressed as a terminating decimal or a repeating decimal. Further exploration may convince you that all rational numbers expressed as simplified fractions will terminate if, and only if, all factors of the denominator are powers of 2 or powers of 5, or a combination of these.

All rational numbers can be expressed as decimals, but is the converse true? Consider the question in Reflect 1.14.

Reflect 1.14

Can all decimals be expressed as fractions?

Any terminating decimal, such as 8.427, is easily converted to fraction form ($^{8427}/_{1000}$). Every repeating decimal also represents a number that can be expressed in fraction form. Figure 1.16 shows a way to find a fractional equivalent for 0.333.... The ideas in this example can be applied to find an equivalent fraction for *every* positive repeating decimal. In fact, one of the more interesting decimals to try is 0.9999..., which results in a fraction that might surprise you!

Let	x	=	0.3333...
	$10x$	=	3.3333...
	$10x$	=	3.3333...
	$-1x$		$-0.3333...$
	$9x$	=	3
	x	=	3/9

Fig. 1.16. Converting a repeating decimal into fraction form

We have shown that all fractions can be expressed as terminating or repeating decimals, and vice versa. There is another kind of decimal, appropriately called a *non-repeating* (and *non-terminating*) decimal. These decimals represent irrational numbers, which cannot be expressed as fractions. (See chapter 2 for a further discussion of irrational numbers.) So the answer to the question posed in Reflect 1.14 must be qualified to indicate what kind of decimals are under consideration.

Our discussion of Big Idea 3 has focused on understanding equivalence among various forms of symbolic rational numbers. The idea of equivalence is foundational to making sense of operations with rational numbers, as we shall see when we consider Big Idea 4.

Big Idea 3
Any rational number can be represented in infinitely many equivalent symbolic forms.

Big Idea 4
Computation with rational numbers is an extension of computation with whole numbers but introduces some new ideas and processes.

Computing with Rational Numbers: Big Idea 4

Big Idea 4. *Computation with rational numbers is an extension of computation with whole numbers but introduces some new ideas and processes.*

For the most part, our interpretations of the four arithmetic operations on rational numbers are the same as with whole numbers, but we need to adapt some of them to make sense. In the discussion that follows, we "unpack" Big Idea 4 to show how those interpretations lead to computational strategies and algorithms for fractions and decimals. We provide examples, as well as visual and concrete representations, to help explain problems that may seem abstract and arbitrary in their symbolic forms. The examples are not intended to demonstrate "the correct way" of performing the calculations and estimations; many problems lend themselves to numerous effective strategies. Instead, the purpose is to develop an understanding of the traditional calculation techniques for rational numbers.

Many aspects of rational number computation mirror whole number arithmetic. Parallels to the properties that govern operations on whole numbers—associativity of addition and multiplication, commutativity of addition and multiplication, distributivity of multiplication over addition, and 0 and 1 as the identities for addition and multiplication, respectively—hold for rational numbers. The first essential understanding related to Big Idea 4 addresses other aspects of the operations on rational numbers, and the second focuses on estimation and mental math.

Interpreting rational number operations and algorithms

Essential Understanding 4a. *The interpretations of the operations on rational numbers are essentially the same as those on whole numbers, but some interpretations require adaptation, and the algorithms are different.*

When children begin working with whole numbers, the first operation that they typically use is addition, quickly followed by subtraction. Later work with rational numbers usually follows suit, making addition and subtraction a natural starting point for our discussion of operations on rational numbers.

Addition and subtraction

The interpretations of the two operations addition and subtraction do not change when the numbers become more complicated. Addition can still mean *combining,* as in a recipe calling for $1\,^1/_2$ cups of grated cheddar cheese mixed with $^3/_4$ cup of Swiss. Subtraction can still mean *taking away* (Jean had \$4.37 and bought a bottle of juice for \$1.75), *comparing* (a 2.6-mile versus a 3.4-mile run), or *finding a missing addend* (a particular stock opened the day at $12\,^3/_8$ and closed at $13\,^1/_4$).

To explore the process of adding fractions (subtraction uses similar reasoning), let's first consider the relatively simple case of two fractions that have the same denominator and are based on the same unit. These can be called *similar fractions* (comparable to *like terms* in algebra, such as $2x$ and $3x$). Suppose that we want to add $^2/_7$ and $^3/_7$. As reflected in Essential Understanding 2c, one interpretation of the number $^2/_7$ is as a quantity composed of two copies of the unit fraction $^1/_7$. So combining 2 sevenths with 3 sevenths produces 5 sevenths, just as combining 2 x's with 3 x's results in 5 x's, or stated another way, $2x + 3x = 5x$. (These sums can be illustrated formally by using the distributive property of multiplication over addition for rational numbers or algebraic expressions.) Figure 1.17 depicts this sum.

Essential ← Understanding 2c

One interpretation of a rational number is as a measure.

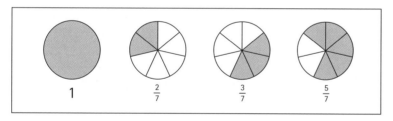

Fig. 1.17. A fraction sum with the same denominator:
$^2/_7 + {}^3/_7 = {}^{2+3}/_7 = {}^5/_7$

Fraction language helps give meaning to this process. On the one hand, the denominator of a fraction can be thought of as naming *what kind* of fractional pieces the fraction has—that is, what unit fraction it uses. (An analogy might be the *denomination* of a currency bill; a \$5 bill and a \$20 bill are different kinds of bills.) The numerator, on the other hand, *enumerates* or counts *how many* of those pieces the fraction has. So, to add two similar fractions like $^2/_7$ and $^3/_7$, we add the numerators to get the total count, while the denominator—the kind of unit fraction the fraction consists of— remains the same.

Unfortunately, when we try to use the same reasoning to add fractions with different denominators, such as $^2/_7$ and $^3/_5$, the process fails. In this case, we are combining 2 of something (the unit fraction $^1/_7$) with 3 of something (the unit fraction $^1/_5$), so we ought to get 5 of something—but what? Not 5 $^1/_7$'s, and not 5 $^1/_5$'s. Figure 1.18 illustrates the dilemma. Because we no longer have equal portions, we don't (yet) know how to label the result with a single fraction. How can we quantify the exact portion represented by the combination $^2/_7 + ^3/_5$?

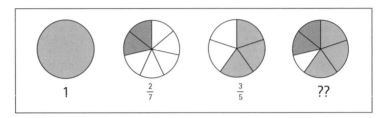

Fig. 1.18. Fraction sum with different denominators: $^2/_7 + ^3/_5 = ?$

Rather than insist on naming the sum as $^5/_{\text{something}}$, we use a different strategy. If we could express both of the fractions that we are adding in forms with the same denominator, we would know how to combine them. This process can be demonstrated with various models; we will use rectangular areas. Figure 1.19 depicts a rectangular unit, with $^2/_7$ of that unit shaded vertically and $^3/_5$ of the same unit shaded horizontally.

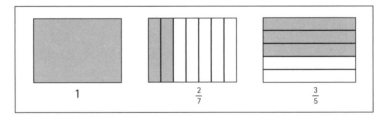

Fig. 1.19. Fraction addition, modeling the numbers: $^2/_7 + ^3/_5$

To create equal-sized subdivisions in the rectangular units showing $^2/_7$ and $^3/_5$, we simply divide the second rectangle in figure 1.19 horizontally into fifths and the third rectangle vertically into sevenths. Figure 1.20 shows this new division of each of these rectangles in the same way that its partner was divided originally.

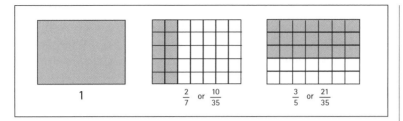

Fig. 1.20. Fraction addition, finding a common denominator:

$^2/_7 + {}^3/_5 = {}^{10}/_{35} + {}^{21}/_{35}$

The result is two fractions that are equivalent to the originals but with a common denominator. The total shaded area, expressed as a fraction of the same unit rectangle, is $^{31}/_{35}$, as shown in figure 1.21.

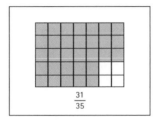

Fig. 1.21. Fraction addition, completing the solution:

$^2/_7 + {}^3/_5 = {}^{10}/_{35} + {}^{21}/_{35} = {}^{31}/_{35}$

Drawing the diagrams in this way—beginning with identical rectangular units that show, respectively, one of the two fractional addends as horizontal shaded bars and the other addend as vertical shaded bars—automatically produces a common denominator (though not necessarily the smallest one) and mirrors the steps in the standard symbolic procedure for adding two fractions. This process conforms to the formal definition of fraction addition:

The sum $\dfrac{a}{b} + \dfrac{c}{d}$ is equal to $\dfrac{ad+bc}{bd}$.

The example above achieves this result through a series of steps:

$$\frac{2}{7} + \frac{3}{5} = \frac{2}{7} \cdot \frac{5}{5} + \frac{3}{5} \cdot \frac{7}{7} = \frac{2 \cdot 5}{7 \cdot 5} + \frac{3 \cdot 7}{5 \cdot 7} = \frac{2 \cdot 5}{7 \cdot 5} + \frac{7 \cdot 3}{7 \cdot 5} = \frac{2 \cdot 5 + 7 \cdot 3}{7 \cdot 5}$$

Note that some steps assume familiarity with fraction multiplication.

We remarked parenthetically in the previous paragraph that the common denominator produced by this process is not necessarily the lowest one. Reflect 1.15 explores whether or not this fact is important in the addition and subtraction of fractions.

Reflect 1.15

Do you have to use the lowest common denominator when adding or subtracting fractions?

Why or why not?

When the denominators of two or more fractions share a common factor greater than 1, as in $3/4$ and $5/6$, we can always find a common denominator (in this case, 12) that is smaller than the product of the denominators (24). Finding a smaller denominator can be more efficient but also more complicated. The essential idea is that *a* common denominator is necessary for adding or subtracting fractions. It doesn't have to be the lowest common denominator; any common denominator will do.

This idea of using any common denominator is helpful when we add decimals. Most of us learned that we can *just line up the decimal points* when we add or subtract decimals such as 32.4 and 6.17. The reason that this rule works is that it also lines up digits corresponding to the same place value (hundredths with hundredths, tenths with tenths, ones with ones, tens with tens, and so on), as illustrated in figure 1.22. Because each place value represents fractions with a fixed denominator ($1/10$'s, for instance), adding the digits in that column is equivalent to adding similar fractions (for example, $4/10 + 1/10$), and placing the sum in that same column (regrouping if necessary) identifies its value appropriately ($5/10$).

Tens	Ones	.	Tenths	Hundredths
3	2	.	4	
	6	.	1	7
3	8	.	5	7

Fig. 1.22. Decimal addition of 32.4 + 6.17

Another interpretation of the standard algorithm for decimal addition merits discussion. Considered in its entirety, the number 32.4 means $324/10$, whereas 6.17 means $617/100$. We can easily create similar fractions by changing the first addend to the equivalent form $^{3240}/_{100}$. Now we can combine *similar decimals* (decimals having the same number of digits to the right of the decimal point), as in figure 1.23. Then aligning the columns starting from the right, as in whole number addition, again permits us to combine digits having the same place value, with the decimal point indicating what each part is worth.

$$\begin{array}{r} 32.40 \\ + \ 6.17 \\ \hline 38.57 \end{array}$$

Fig. 1.23. Decimal addition with similar decimals

Multiplication

Rectangular shapes provide a common model for the product of two whole numbers, showing the product either as the number of elements in a rectangular array (such as 3 rows of 5 desks each) or as the area of a rectangle (of width 3 and length 5). A rectangular array involving rational numbers may or may not be meaningful, depending on the context. In the desk example, $3\,^1/_2$ rows of $5\,^1/_3$ desks each makes no sense. But $^2/_3 \times ^3/_5$, for instance, could be interpreted as an array consisting of $^2/_3$ of a column by $^3/_5$ of a row, as represented by the shaded circles in figure 1.24a. (Of course, we could also call this $^3/_5$ of a row by $^2/_3$ of a column, illustrating the commutativity of rational numbers.) If we consider a continuous area rather than a discrete array of objects, then we are no longer counting items but measuring lengths and areas, which allows us to subdivide any unit into arbitrary fractional parts. Figure 1.24b shows a *unit square* (one unit long, one unit high, and one square unit in area) and a shaded rectangle whose area also represents the product $^2/_3 \times ^3/_5$.

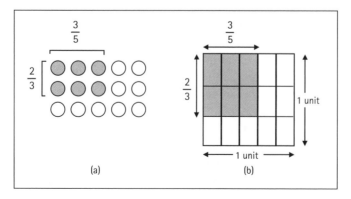

Fig. 1.24. Two models for fraction multiplication: $^2/_3 \times ^3/_5$

These diagrams nicely illustrate the algorithm for multiplying two fractions. In both cases, the two-dimensional unit is partitioned in one dimension into thirds and in the other into fifths, creating fifteenths; that is, the resulting fractional part is the product of the denominators. The parts that we are counting also form a rectangular shape representing the product of the numerators. In symbols,

$$\frac{2}{3} \times \frac{3}{5} = \frac{2\times3}{3\times5},$$

which is easily seen to be 6 out of 15 equal parts, or $^6/_{15}$. (The example provides us with a case in which the simplest fractional form, $^2/_5$, is less useful than $^6/_{15}$ for our purpose, which is to explain the process.) Both diagrams provide visual models for the formal definition of fraction multiplication:

The product $\dfrac{a}{b} \times \dfrac{c}{d}$ is equal to $\dfrac{ac}{bd}$.

In the previous discussion of modeling fraction multiplication, we used fractions abstractly, without reference to quantities in actual situations. Reflect 1.16 turns the focus from the abstract to the concrete.

Reflect 1.16

Can you think of a "real" context for $^2/_3 \times {}^3/_5$ that you could model with the diagram in figure 1.24a? How about with the diagram in figure 1.24b?

In each case, what would you be representing by each part of the diagram (a single small circle, the entire set of 15 circles, one of the smallest rectangles, the largest rectangle)? Which quantities would play the role of the unit?

How would your two situations be the same, and how would they be different?

If possible, compare your ideas about Reflect 1.16 with ideas from your colleagues. Thinking about different ideas and examples may sharpen your insight into the distinction between discrete and continuous quantities and models.

The operator or scalar interpretation is a powerful approach to whole number multiplication (see Otto et al. [forthcoming]). Using this interpretation, we can think of the product 3 × 5 as the quantity that is 3 times as great as 5. One common way of expressing this quantity is as the familiar "repeated addition" 5 + 5 + 5, which is easily modeled on the number line, as shown in figure 1.25. (Some authors say that 3 × 5 means 5 groups or copies of 3, or 3 + 3 + 3 + 3 + 3. Because multiplication of rational numbers is commutative, both interpretations give rise to the same result and are reasonable, but 5 + 5 + 5 is more consistent with the operator approach.)

Developing Essential Understanding of Multiplication and Division for Teaching Mathematics in Grades 3–5 (Otto et al., forthcoming) presents the operator, or scalar, interpretation of whole number multiplication as a powerful approach.

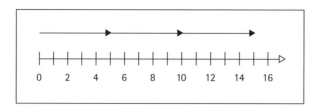

Fig. 1.25. Repeated addition: 3 × 5 = 5 + 5 + 5, or 15

This approach extends easily to rational numbers when the operator (the multiplier) is a whole number. Figure 1.26 illustrates the product $3 \times \frac{3}{5}$ as three copies of $\frac{3}{5}$ on the number line.

Fig. 1.26. Repeated addition: $3 \times \frac{3}{5} = \frac{3}{5} + \frac{3}{5} + \frac{3}{5}$, or $\frac{9}{5}$, or $1\frac{4}{5}$

When the multiplier is a fraction, however, as in $\frac{2}{3} \times \frac{3}{5}$, the operator approach provides a more meaningful model than repeated addition. If one pound of cheese costs \$5, then 3 pounds cost \$15. If one pound of flour costs \$$\frac{3}{5}$, then 3 pounds cost \$$\frac{9}{5}$. And $\frac{2}{3}$ of a pound of flour costs $\frac{2}{3}$ of \$$\frac{3}{5}$, or \$$\frac{2}{5}$. The product $\frac{2}{3} \times \frac{3}{5}$ signifies a quantity that is $\frac{2}{3}$ as big as $\frac{3}{5}$, which we often describe simply as "$\frac{2}{3}$ of $\frac{3}{5}$." This formulation gives us another way to think about the two diagrams in figure 1.24: to model the fraction multiplication $\frac{2}{3} \times \frac{3}{5}$, we can begin with a unit whole, take $\frac{3}{5}$ of it, and then take $\frac{2}{3}$ of that.

Alternatively, adapting the number line model to $\frac{2}{3} \times \frac{3}{5}$ produces figure 1.27. Note that in this particular instance, the $\frac{3}{5}$ length is already divided into three equal segments, so that we can find $\frac{2}{3}$ of it directly without introducing further subdivisions. Different numbers may necessitate an additional step.

Fig. 1.27. A fraction of a fraction: $\frac{2}{3} \times \frac{3}{5} = \frac{2}{3}$ of $\frac{3}{5}$, or $\frac{2}{5}$

Does this understanding of fraction multiplication help in multiplying decimals? Reflect 1.17 investigates an aspect of the standard procedure for decimal multiplication.

Reflect 1.17

The standard algorithm for multiplying two decimal forms is often memorized as a rule telling how many decimal places to put in the answer.

Why does this rule work?

The standard procedure for decimal multiplication follows directly from the way in which we multiply ordinary fractions. For example, 0.27×1.5 can be rewritten as $^{27}/_{100} \times {}^{15}/_{10}$. The product is

$$\frac{27 \times 15}{100 \times 10}, \text{ or } \frac{27 \times 15}{1000}.$$

We can change this product back into decimal form by multiplying 27 and 15 as whole numbers to get 405, then indicating thousandths with three decimal places, giving 0.405 as the answer. More generally, terminating decimals can be written as fractions with denominators that are powers of 10 and numerators that are integers. When two such denominators are multiplied, the product is again a power of 10. In the example above, $100 \times 10 = 10^2 \times 10^1$, and the result is 1000, or 10^3. The exponent of the product is the sum of the exponents of the factors. The multiplication algorithm encodes this as, "The number of decimal places in the answer is the sum of the numbers of decimal places in the two numbers being multiplied."

Division

Most people have little understanding of what division means when fractions are involved. We'll begin with the more familiar case of whole number division, which lends itself to two different interpretations, and then see how this case can be adapted to the division of fractions.

If 20 ounces of juice are shared equally among 4 children, how much juice does each child receive? This problem exemplifies the *partitive* interpretation of the division $20 \div 4$, in which the dividend (the quantity to be divided) is partitioned or separated into a *given number of equal-sized parts*. The quotient, 5 ounces per child, tells how much is in each part. Partitive division is sometimes called "fair sharing."

Suppose that we try to use the partitive interpretation with division involving fractions. If $^2/_3$ of a pint of juice is shared equally among 4 children, how much juice does each child receive? This division $2/_3 \div 4$ can be accomplished easily, especially if we think of $2/_3$ of a pint as $4/_6$ of a pint. Then each portion is $1/_6$ of a pint. Now let's see what happens when we reverse the roles of the numbers. When 4 pints of juice are shared equally among $2/_3$ of a person... Oops! Our reversed scenario is absurd. Fair sharing doesn't really fit when the divisor is not a whole number.

Partitive division lends itself to another interpretation for fractional divisors, however. Again, we will begin with a straightforward whole number problem. If a swimmer drinks 20 pints of juice at a steady rate over 4 days, what is her daily consumption? The quotient $20 \div 4$ equals 5 pints per day. This quotient can be thought of as a "per unit" quantity resulting from partitioning the juice into 4 equal-sized daily portions. Now consider the parallel situation of a runner for whom 4 pints of juice represents $2/_3$ of his daily allotment (that is, he drinks 4 pints in $2/_3$ of a day); what is his daily consumption? His per-unit rate is $4 \div 2/_3$ pints of juice per day.

We can make sense of these situations as division problems, but how can we actually carry out the operation $4 \div 2/_3$ and express the result as a single number? A rule that focuses on procedures to the exclusion of sense making is the old standby: "Ours is not to reason why; just invert and multiply." But let's flout convention and try to reason through this rule and discover why it works.

One justification for "invert and multiply" draws on the definition of division, which is based on its inverse relationship with multiplication. We examined the idea that every division problem can be rewritten as a multiplication problem in our discussion of Essential Understanding 1a. For example, sharing 20 ounces of juice equally among 4 children results in 5 ounces of juice per child, because if 4 children each drink 5 ounces of juice, they will consume 20 ounces all together. In symbols, $20 \div 4 = 5$ because $5 \times 4 = 20$. (Order is sometimes a source of confusion. In the juice context, 4 is the multiplier [operator], so this product would be written as $4 \times 5 = 20$. As the inverse of the division $20 \div 4 = 5$, it is more properly expressed as $5 \times 4 = 20$. Commutativity assures us that the two products are equivalent.)

We see that applying the same reasoning to the division sentence $4 \div 2/_3 = \square$ leads to the multiplication sentence $\square \times 2/_3 = 4$. Now we need to use the property that holds that multiplying any rational number (except 0) by its *multiplicative inverse* (its *reciprocal*) results in the *multiplicative identity* 1. Starting with the original division equation and keeping it balanced, we have:

Essential ← Understanding 1a

Rational numbers are a natural extension of the way that we use numbers.

Recall from the discussion of Essential Understanding 1a that every division problem of the form $a/_b = \square$ can be rewritten as $\square \times b = a$ (see p. 15).

$$4 \div \frac{2}{3} = \square$$

$$\square \times \frac{2}{3} = 4$$

$$\square \times \frac{2}{3} \times \frac{3}{2} = 4 \times \frac{3}{2}$$

$$\square \times 1 = 4 \times \frac{3}{2}$$

$$\square = 4 \times \frac{3}{2}$$

That is, $4 \div \frac{2}{3} = 4 \times \frac{3}{2}$, or 6, since both expressions are equal to the number represented by the box. In context, if the runner drinks 6 pints of juice a day, then $\frac{2}{3}$ of his daily ration is 4 pints. Because $\frac{2}{3} \times 6 = 4$, we conclude that $4 \div \frac{2}{3} = 6$. Looking at the numbers only, we see that dividing 4 by the fraction $\frac{2}{3}$ is equivalent to multiplying 4 by the reciprocal fraction $\frac{3}{2}$. Repeating the same series of steps more deliberately while using variables instead of numbers would lead to a proof that dividing a number by a fractional divisor has the same effect as multiplying that number by the reciprocal of the divisor. Reflect 1.18 explores these ideas further.

Reflect 1.18

For the division problem $\frac{2}{5} \div \frac{3}{4}$, would the multiplication $\frac{5}{2} \times \frac{3}{4}$ give the correct quotient?

If not, what would it produce? Why?

We've established that $\frac{2}{5} \div \frac{3}{4}$ has the same value as $\frac{2}{5} \times \frac{4}{3}$, or $\frac{8}{15}$. The product $\frac{5}{2} \times \frac{3}{4}$ is equivalent to $\frac{15}{8}$, the reciprocal of $\frac{8}{15}$. Is this a coincidence? Consider a simpler case: $6 \div 2$ is 3, whereas $2 \div 6$ is $\frac{1}{3}$. You might explore other cases or other lines of reasoning to establish the relationship between $a \div b$ and $b \div a$ for rational number division in general and the effect of inverting the first, versus the second, fraction.

The other basic interpretation of whole number division is *quotative*. If 20 ounces of juice are poured into 4-ounce glasses, how many glassfuls will there be? The arithmetic operation is the same as in the first partitive example, $20 \div 4$, but in quotative division, the dividend is separated into *parts of a given size*. In this case, the quotient, 5 glassfuls, describes how many parts result.

What happens when we try to apply the quotative interpretation to $4 \div \frac{2}{3}$? An example might entail pouring 4 pints of juice into glasses holding $\frac{2}{3}$ of a pint; how many glassfuls will result? This question is reasonable. Moreover, the way in which someone would actually carry out the pouring leads directly to the answer through the process of repeated subtraction, which is one way of modeling quotative division. The pourer would begin with 4 pints of juice, and then fill one $^2/_3$-pint glass; $3\,^1/_3$ pints would remain. Then the pourer would fill a second glass, and $2\,^2/_3$ pints would be left; filling a third glass would leave 2 pints. Continuing to pour three more times would use up all the juice, with 6 "pourings" altogether. As this example suggests, quotative division, or separating some quantity into parts of a given size, can be interpreted in the same way for fractional divisors as for whole number divisors.

The quotative interpretation of division is often called the *measurement interpretation,* referring to another way of modeling the process. As usual, it makes sense to begin with what we know about working with whole numbers. The quotative interpretation of 12 ÷ 3, for example, says to divide 12 into parts of size 3, then count how many parts you have. A simple way to frame the question is, "How many 3s are in 12?" This language lends itself to a measurement interpretation, as in "How many 3-foot yardsticks does it take to make up a length of 12 feet?" The answer is easily found by actually carrying out the measurement.

The concept of unit plays a central role in this measurement scenario. The number 12 describes a length with respect to the unit of length 1 foot (a ruler). The quotient 12 ÷ 3 can be interpreted as describing that same length (12 feet) with respect to a new unit of 3 feet (a yardstick). We can think of the divisor, 3, as a new measuring rod or unit length for measuring 12. The quotient, 4, is the number of units of 3 feet that are in 12 feet.

This measurement interpretation of division is especially helpful when the quotient is not a whole number, and it works well with division of fractions. The diagrams in figure 1.28 illustrate the process for $2\,^1/_3 \div\,^5/_6$. Note that both $2\,^1/_3$ and $^5/_6$ are defined in terms of the original unit (1). The question asked by the division $2\,^1/_3 \div\,^5/_6$ is, "How many copies of the new unit (the $^5/_6$ measuring rod) make up $2\,^1/_3$?"

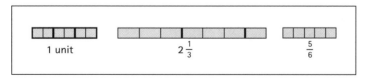

1 unit $2\frac{1}{3}$ $\frac{5}{6}$

Fig. 1.28. Fraction division, modeling the numbers in $2\,^1/_3 \div\,^5/_6$

Figure 1.29 shows that $2\,^1/_3$ contains more than 2, but a little less than 3, copies of the $^5/_6$ rod. How much more than 2? The diagram indicates that $2\,^1/_3$ contains $^4/_5$ of another copy of the $^5/_6$ rod, so the total measurement is $2\,^4/_5$ copies of the new measuring rod.

In other words, it takes $2\,^4/_5$ copies of $^5/_6$ to make $2\,^1/_3$, which means that $2\,^4/_5 \times\, ^5/_6 = 2\,^1/_3$. Consequently, $2\,^1/_3 \div\, ^5/_6 = 2\,^4/_5$.

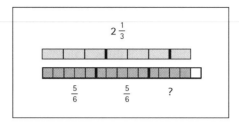

Fig. 1.29. Fraction division using the divisor as a measuring rod in $2\,^1/_3 \div\, ^5/_6$

A common error in this process is to interpret the portion (the four small squares) of the last measuring rod as $^4/_6$, since they were equivalent to $^4/_6$ of the *original* unit. But now they are being considered as a fraction of the *new* unit $^5/_6$—that is, as a fraction ($^4/_5$) of the divisor ($^5/_6$).

The measurement model provides one way to challenge the naïve belief, derived from countless whole number examples, that division always makes numbers smaller. Any quantity expressed as a length will have a greater measure with a $^5/_6$ rod than with a unit rod. Similarly, the operator interpretation of multiplication provides a way to challenge the whole number view that multiplication makes numbers bigger; $^2/_3$ of any positive quantity is clearly less than the original quantity.

We've discussed how quotative division is related both to repeated subtraction and to measurement. Reflect 1.19 explores this relationship.

Reflect 1.19

Suppose that you are baking a cake according to a recipe that calls for $1\,^1/_2$ cups of flour, and your only clean measuring cup holds $^1/_3$ of a cup. How many $^1/_3$-cupfuls should you use for your recipe?

Use this problem to explore how the repeated subtraction and measurement approaches to division are related.

Turning to decimal forms, we can see that the standard procedure for dividing follows directly from equivalence of fractions. In the decimal division $0.012\overline{)4.8}$, for instance, the "cookbook" rule says to move the decimal point in the divisor three places to the right, to get a whole number (12), then move the decimal point in the dividend the same number of places to the right, and then go straight up for the decimal point in the quotient. The result is an alternative quotient, $12.\overline{)4800}$, whose value is 400. To see why this rule works, we can write the division $4.8 \div 0.012$ as the fraction $^{4.8}/_{0.012}$. We can then multiply $^{4.8}/_{0.012} \times {}^{1000}/_{1000}$ to change it to the equivalent fraction $^{4800}/_{12}$, for which $12.\overline{)4800}$ could be used to calculate the correct quotient. Moving the two decimal points the same distance in the same direction maintains the same ratio between the dividend (numerator) and the divisor (denominator) because each movement results in multiplication by the same power of 10.

Estimation and mental math with rational numbers

Essential Understanding 4b. *Estimation and mental math are more complex with rational numbers than with whole numbers.*

School math traditionally emphasizes paper-and-pencil computation and exact answers. But adults do in their heads much of the arithmetic that they use in their daily lives, and often an approximate answer suffices for practical purposes. Mental calculation and estimation with rational numbers are essential skills but involve more complex reasoning than with whole numbers, as the everyday example in Reflect 1.20 demonstrates.

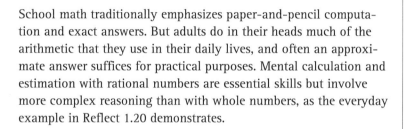

Reflect 1.20

Imagine yourself shopping for groceries. You find two brands of tomatoes, one sold in a 16-ounce can that costs $1.49, and the other sold in a 28-ounce can that costs $2.89.

How do you determine which is the better buy?

What aspects of rational number sense might come into play?

Assuming that quality is not an issue, the smaller dollar-per-ounce ratio is the more economical. Determining which brand has the smaller ratio leads to comparing the two rational numbers $^{1.49}/_{16}$ and $^{2.89}/_{28}$. Converting these forms to decimals would make the comparison easy but is hard to do without a calculator, which many people don't use when shopping. Instead, you might round prices to get the simpler values $^{1.5}/_{16}$ and $^{2.9}/_{28}$. Then you could easily see that both are close to the benchmark fraction $^{1}/_{10}$, but the first is smaller than $^{1}/_{10}$ and the second is larger than $^{1}/_{10}$, thus solving the problem. Of course, other paths to the solution are possible with different strategies.

Consider what solving this particular problem entails. Given two whole numbers, we can easily identify the smaller by using an understanding of place value and the counting sequence. With two fractions, however, identifying the smaller is usually not so straightforward, as this example shows. With fractions, we often rely on benchmarks for identifying approximate magnitudes, and we draw on an understanding of equivalence for making comparisons when the denominators are different. Rounding 2.89 to 2.9 involves a sophisticated understanding of decimal place value. Recognizing that $^{1.5}/_{16}$ and $^{2.9}/_{28}$ are both close to $^1/_{10}$ is founded in decimal-fraction equivalence as well as a familiarity with benchmark fractions. In short, mentally comparing whole numbers, such as 238 and 253, is much less demanding than comparing fraction forms, like $^{1.49}/_{16}$ and $^{2.89}/_{28}$, or even $^5/_9$ and $^7/_{12}$.

Decimals are similar in form to whole numbers but can be more challenging to estimate. Reflect 1.21 explores an example.

Reflect 1.21

Suppose that you use a calculator to find the product 0.427 × 0.00652, and the display shows the number 0.0278404.

Do you trust this answer? Explain.

Estimation is important even when we are using a calculator. Although the device usually gives nearly exact answers, it is easy to push a wrong button; checking the approximate result can identify an error. In this case, recognizing that 0.427 is a little less than $^1/_2$ indicates that the product should be close to 0.003, not 0.03. This reasoning, as in the previous example, entails rational number sense, manifested as an understanding of, and fluency in using, equivalence between different numerical forms, approximation, benchmarks, and rational number operations.

Conclusion

Big Idea 4
Computation with rational numbers is an extension of computation with whole numbers but introduces some new ideas and processes.

The examples that we have presented throughout our discussion of Big Idea 4 suggest the complexity inherent in calculation with rational numbers, and the value of visual and contextual models in explaining the procedures. Whether symbolic or pictorial or manipulative, written down or acted out or thought through, computing with rational numbers involves more complicated reasoning than computing exclusively with whole numbers does.

In fact, *complexity* is a theme that runs through all four big ideas in this chapter. Fractions are numbers that are essentially quotients of whole numbers, and this idea in itself can be confusing. The payoff for understanding it is an awareness of the fact that fractions and rational numbers extend the whole numbers in powerful ways and enable us to solve otherwise unsolvable problems (Big Idea 1). Grasping the multiple interpretations of rational numbers presents another challenge, but making sense of each interpretation builds insight into numerous areas of mathematics (Big Idea 2). Although every whole number corresponds to a single numeral, every rational number can be written in many different ways. The compensation for becoming skillful in dealing with this additional complexity is having an infinite number of forms of a single number and access to particular forms that may be more suitable than others for a given purpose (Big Idea 3). Performing calculations with rational numbers is more complicated than with whole numbers, but an understanding of rational numbers and arithmetic operations leads to computational procedures for rational numbers that are extremely valuable and make sense (Big Idea 4).

Ideas associated with rational numbers are used throughout mathematics. Chapter 2 describes some of these connections.

Big Idea 1
Extending from whole numbers to rational numbers creates a more powerful and complicated number system.

Big Idea 2
Rational numbers have multiple interpretations, and making sense of them depends on identifying the unit.

Big Idea 3
Any rational number can be represented in infinitely many equivalent symbolic forms.

Connections: Looking Back and Ahead in Learning

Rational numbers are an important topic in grades 3–5, but like other topics in the mathematics curriculum, they do not occur in isolation. Moreover, they build on ideas and understandings from the primary grades and continue to develop throughout the grades. This chapter shows how key ideas about whole numbers learned in earlier grades underpin the essential understandings about rational numbers identified in chapter 1. It also shows how mathematics topics in higher grades extend and apply the ideas about rational numbers that students begin developing in grades 3–5, and how these ideas gradually lead to number systems beyond rational numbers.

Building from Whole Numbers to Rational Numbers

Conceptual bases for multiplication and division of whole numbers, base-ten place value, and the properties of whole number operations all serve as foundations for understanding rational numbers. Students are still developing these concepts as they encounter new mathematical ideas associated with rational numbers.

Counting, unitizing, and multiplication

Counting *groups* of objects is much more difficult for children to learn than simply counting a set of discrete objects. Counting groups and learning multiplication require *unitizing*—that is, conceptualizing a group of several objects as a single entity. Unitizing is a key building block for working with rational numbers because a set of discrete objects must be considered as a unit before it can be apportioned into equal-sized groups. Just as 3×6 can be understood as 3 groups of 6 objects, so too can $1/3 \times 6$ be seen as one of three equal-sized subgroups of 6 objects. Sometimes, however, key concepts from whole number arithmetic interfere with rational number sense. For example, when operating exclusively with counting numbers greater than 1, multiplication always results in products that are larger than either factor. But when we are multiplying a positive number by a number between 0 and 1, we discover that the product is less than the larger factor.

Fractions and division

The intuitive idea of "fair shares" forms a conceptual basis for whole number division. It also serves as a building block for fractions, where the denominator names some number of equivalent parts. Essential Understanding 2d captures the idea that one of the interpretations of a fraction (or of any rational number) is as a quotient, emphasizing the division of the numerator by the denominator. In division, the remainders, too, can be related to fractions. For example, when 2 children try to share 7 markers fairly, directly modeling the problem leaves 1 marker remaining. This quotient doesn't make sense as $3\frac{1}{2}$ or 3.5; 3R1 is a meaningful alternative. But if those same 2 children are sharing 7 feet of yarn, the practical solution could reasonably be expressed as $3\frac{1}{2}$ feet (see the discussion of Essential Understanding 2d).

On the basis of early work in dividing with whole numbers, students commonly think that the quotient is always less than the dividend. Later, however, when working with fractions, they discover that the quotient is *not* necessarily less than the dividend, as discussed in Essential Understanding 4a. Although concepts developed in whole number arithmetic form a useful base for understanding operations with rational numbers, misconceptions can arise from overgeneralizing.

Decimals and place value

Our relatively complex numeration system for whole numbers is built on a few basic properties, including place values representing powers of 10. Learning how this system works is a central topic of study in the early grades. Soon afterward, students encounter rational numbers in the form of decimals. Exactly the same principles govern the construction of decimals as of whole numbers, though some of the patterns must be extended to represent fractional amounts. In our discussion of Essential Understanding 3c, we described whole number numeration in detail and then showed how it forms the basis for decimals.

→ Essential
Understanding 2d

*One interpretation
of a rational number
is as a quotient.*

→ Essential
Understanding 4a

*The interpretations
of the operations on
rational numbers
are essentially the
same as those on
whole numbers, but
some interpretations
require adaptation,
and the algorithms
are different.*

→ Essential
Understanding 3c

*A rational number
can be expressed as
a decimal.*

Using Rational Numbers across the Curriculum

Whole numbers are fully satisfactory for work with many aspects of school mathematics and in many everyday situations outside the classroom. However, in a myriad of other circumstances, they are woefully inadequate. These situations frequently involve topics such as measurement, ratios and proportions, percents, probability and data analysis, and algebra.

Measurement

Sometimes whole numbers are practical and perfectly adequate for measuring. In cutting and sewing window curtains, for instance, measuring the fabric to the nearest inch is probably fine. But to re-place a piece of window glass, cutting the glass to $30'' \times 46''$ won't work if the frame is really $30\frac{1}{4}'' \times 45\frac{5}{8}''$. Students in grades 3–5 learn to make more and more precise measurements of various attributes, creating a need for fractions and decimals, together with reasonable approximations. Later they learn that although $\frac{1}{2}$ and $\frac{4}{8}$ may be equivalent fractions, they can represent different degrees of precision when used in a measurement context. A measurement reported as $\frac{1}{2}''$ could represent any length between $\frac{1}{4}''$ and $\frac{3}{4}''$, but a measurement of $\frac{4}{8}''$ designates a length between $\frac{7}{16}''$ and $\frac{9}{16}''$ (that is, closer to $\frac{4}{8}''$ than to $\frac{3}{8}''$ or $\frac{5}{8}''$).

In science applications, students extend the idea of precision to significant digits. Scientists and technicians typically use rational numbers in decimal form and include fewer or more decimal digits to express measurements to whatever degree of precision is justified and useful. A physicist who wants red light might generate a frequency of 0.00000064 meters, whereas light with a frequency of 0.00000056 meters is green. Rounding either of these values to 0.0000006 meters would be inappropriate, since this is the frequency of orange light.

Ratios and proportional reasoning

Ratio and proportionality are central topics in middle school mathematics. Ratios use ideas related to rational numbers, but they are not identical to fractions. As discussed in connection with Essential Understanding 2e, a ratio is a multiplicative comparison of two quantities. A proportion is an equation stating that two ratios are equivalent. Reasoning with proportions is the fundamental basis for many ideas and applications in mathematics and other quantitative sciences.

 Essential Understanding 2e

One interpretation of a rational number is as a ratio.

Developing Essential Understanding of Ratios, Proportions, and Proportional Reasoning for Teaching Mathematics in Grades 6–8 (Lobato and Ellis 2010) presents a full discussion of proportionality in mathematics and everyday life.

For an extended investigation of the use of proportions in marine biology to count fish, see the activity How Many Fish in the Pond? in *Navigating through Mathematical Connections in Grades 6–8* (Pugalee et al. 2008, pp. 88–97, 145–49).

For example, suppose that a marine biologist captures 63 fish in a small lake, tags them, and then releases them. If she captures 47 fish a month later and finds that 8 of them are tagged, she can then solve the proportion $^8/_{47} = {}^{63}/_?$ to estimate that there are 370 fish in the lake. Proportions are common in everyday life, too. Suppose that a recipe for lemonade syrup for 50 people calls for 4 cups of water, 8 cups of sugar, and the juice of 25 lemons. The home cook who adapts this recipe to serve 5 or 10 or 20 people is using proportions.

Percents

Sometimes it is convenient to express rational numbers representing ratios in percent form, as in the example of comparing numbers of successful shots out of free-throw attempts in chapter 1's discussion of Essential Understanding 2e (see p. 26). Both scientific reports and the popular media often use the language of percents to describe numerical data. For instance, consider the circle graph in figure 2.1. A reporter might use such a graph if he wanted to help the public understand that at a particular emergency relief agency, more than 50% of the expenditures had gone into salaries, while less than 20% of the expenditures had actually gone to direct aid to disaster victims.

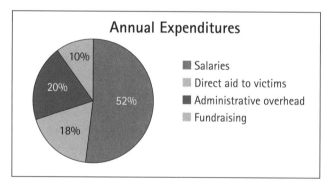

Fig. 2.1. A circle graph can represent parts of a whole as percentages

A circle graph is similar to a "pie" representation for fractions and may be used to provide a visual representation of a whole (unit) and its parts. Note, however, that a circle graph is typically divided into parts that are *not* the same size. When percents are used to describe part-whole situations, they can be added together as long as they are all part of the same whole. So the total percent of expenditures shown in the circle graph can be found by adding:

$$52\% + 18\% + 20\% + 10\% = 100\%.$$

Percentages like 110% or 0.25% are often poorly understood. To illustrate, think about what 0.25% means in the context of the circle graph in figure 2.1. A common error is to think that 0.25% means 25%. This error would cause someone to think that 0.25% represents $1/4$ of the budget in the circle graph, whereas it actually represents a tiny amount of the budget—just $1/4$ of one percent. By contrast, what would 110% mean in the context of the circle graph? In a typical circle graph, the circle represents a total amount as the unit on which the percents are based. In the graph in figure 2.1, percentages represent the parts of 100% of the expenditures, so 110% would not make sense in this context. Percentages greater than 100, however, can be meaningful and useful in other contexts. For example, suppose that the budget for an agency is $750,000 and its expenditures are $825,000. Then the expenditures are 110% of the budget. Circle graphs are not used for percentages greater than 100, since the convention is to use only one circle and to use the entire circle to represent 100%.

Although percents are sometimes considered an alternate system of symbols for rational numbers, not all interpretations permit fractions or decimals to be replaced sensibly or reasonably by percents. For example, although $75/100$ and 75% represent the same fraction of a whole, we wouldn't replace $.75 with 75% dollars. The quantity 75 *cents* is not the same as 75 *per cent*. Similarly, although $2\,1/2$ is equivalent to 250%, we wouldn't say 250% pizzas for $2\frac{1}{2}$ pizzas. We could, though, easily use either $3/4$ or 75% to describe a portion of a cake.

Experts have long debated whether *percent* indicates a ratio, a number, or both (see Parker and Leinhardt [1995], for example). The term *percent* has various interpretations and can be used in different ways; however, it always resides in the world of multiplicative comparisons. The expression 15% can indicate a part-whole comparison, a ratio (such as 15 of one quantity for every 100 of another—for example, 15 almonds for every 100 peanuts in a nut mix), an interest rate, the probability of an event, or an operator (as in 15% of some quantity). No matter how percents are formally defined, these examples show that they are closely related to rational numbers.

Probability and data analysis

Probability is the mathematics of uncertainty, the science of predicting how likely it is that a particular event will occur. If you have two dimes and six pennies in your pocket and you reach in and pull out one coin randomly, what are the chances that it will be a dime? Because you have eight coins in total and two of the eight are dimes, the probability of choosing a dime is $2/8$ (or $1/4$, or 0.25, or 25%).

Every probability is defined as a part-whole comparison, comparing the number (or the measure) of favorable outcomes in a given situation to the total number (or the total measure) of possible outcomes. The probability of a particular event is a number between 0 and 1, inclusive, ranging from impossible (0) to certain (1). In grades 3–5, students begin using fractions to quantify likelihood, both by gathering data (for example, to predict which way a thumbtack will land when tossed) and by counting theoretically possible outcomes (as with game spinners). In later grades, they will refine these techniques and also discover that some probabilities are not rational numbers.

As students begin collecting a variety of data in grades 3–5, they are faced with the tasks of organizing, representing, analyzing, and interpreting them. As mentioned above, measurement data often come in the form of fractions, decimals, and mixed numbers. Simply ordering a collection of such numbers is a nontrivial task. Even if the data consist solely of whole numbers, computing measures of central tendency—the mean and even the median—may well result in rational numbers. More sophisticated forms of data analysis in later grades rely heavily on rational expressions and their operations and on rational number sense.

Algebra

As students progress through middle school and into high school, algebra becomes a dominant feature of the curriculum. Rational numbers and their operations are essential elements of algebra in numerous ways. Perhaps the most obvious of these is in solving equations. In chapter 1, we showed that the solution to even a simple equation, like $5x = 12$, might be a rational number, such as $^{12}/_5$. If you know that a dollar is currently worth 12.8 pesos, then you can find how many dollars (d) a 500-peso note is worth by solving $12.8d = 500$. Solving and checking an equation like $(^3/_4)x + {}^7/_8 = {}^5/_2$ entails using at least three of the operations of fraction arithmetic.

Fractions containing an algebraic variable are called rational expressions. For instance, the basic formula $d=rt$ (distance equals rate times time) can be solved algebraically to find a fraction representing either a rate $(^d/_t)$ or a time interval $(^d/_r)$, depending on the context, and then applied to solve a variety of problems involving rates. A typical example follows.

Ordinarily, Antonio rides his bike at 15 miles per hour. On a breezy day, he rode 20 miles directly into the wind and then turned around and rode home with the wind. His entire trip took 3 hours. What was the speed of the wind?

An algebraic solution entails solving the equation

$$\frac{20}{15-x} + \frac{20}{15+x} = 3,$$

where x represents the speed of the wind, and reveals the answer as 5 miles per hour. This equation is a mathematical representation of the statement "(time interval traveling against the wind) + (time interval traveling with the wind) = 3 hours, the total time." People often make fun of such contrived problems (this solution makes the unrealistic assumptions that the wind speed is constant and that a 5-mph wind has a 5-mph effect on a cyclist; actually, the effect is much smaller). Some people remember these types of problems as horror stories. Yet, the operations that are required to work with rational expressions and equations obey essentially the same rules as the rational number operations studied in elementary school. Operations with rational expressions extend beyond algebra to trigonometry and calculus as well.

In learning algebra, students extend their study of exponents beyond repeated multiplication to include negative exponents, as in 3^{-2}. They are often surprised to find that these exponents lead to fractions! A negative exponent for a number signals a reciprocal, so that

$$3^{-2} = \frac{1}{3^2}, \text{ or } \frac{1}{9}$$

This idea applies to algebraic expressions, too: for any x different from 0,

$$x^{-n} = \frac{1}{x^n}.$$

Fractions can even be used as exponents; for instance,

$$8^{\frac{2}{3}} \text{ means } \left(\sqrt[3]{8}\right)^2,$$

which simplifies to 4. Fractional exponents are central to the study of logarithms.

Other aspects of algebra, such as proportional and other linear relationships expressed as algebraic equations and graphs, also make use of ratios, rational numbers, and rational expressions (see Lobato and Ellis 2010). The understanding of fractions that children develop in elementary and middle school is put to direct use in many ways when they study algebra.

Developing Essential Understanding of Ratios, Proportions, and Proportional Reasoning for Teaching Mathematics in Grades 6–8 (Lobato and Ellis 2010) offers further discussion of ratios, rational numbers, and rational expressions in proportional relationships represented in algebra.

Extending beyond Rational Numbers

Chapter 1 explained how the rational numbers evolved historically, building on simpler number systems. We also noted that the word *fraction* has various interpretations, and we chose to let it mean *a symbolic expression of the form $^a/_b$ representing a nonnegative rational number,* the interpretation that is prevalent in grades 3–5.

Negatives of fractions

It is important to acknowledge that every positive nonzero fraction has a corresponding negative, such as $^2/_3$ and $-^2/_3$, and that these forms also represent rational numbers. (Some texts use the term *fractions* to mean this larger set of negative and nonnegative rational numbers.) Performing arithmetic operations on negative fractions is just a matter of combining principles about signed numbers together with principles about rational numbers; no special considerations apply.

Irrational and real numbers

→ Essential
Understanding 3*b*

*Between any two
rational numbers
there are infinitely
many rational
numbers.*

Essential Understanding 3*b* expresses the idea that between any two rational numbers there exist infinitely many more rational numbers. If we magnify the number line, as illustrated in figure 2.2, it would appear that the number line is "full."

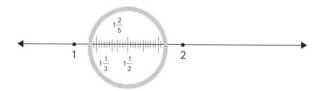

Fig. 2.2. Magnifying the number line

It may be hard to imagine that although the number line now appears to be "full" with points corresponding to all of the rational numbers, it actually contains an even greater collection of points that do *not* correspond to rational numbers! That is, the rational number line has lots of "holes," even though it appears solid.

The previous chapter discussed another way of defining a rational number—as any number whose decimal expansion results in either a terminating or a repeating decimal. Examples include $^1/_8$ = 0.125; $^1/_3$ = 0.33333...; $^3/_7$ = 0.428571428571...; and $^{312,161}/_{99,900}$ = 3.12473473473.... Numbers whose decimal expansions do not terminate or form a repeating decimal pattern are called *irrational numbers*. To put it another way, we can say that it's impossible to find integers *a* and *b* that will make the fraction $^a/_b$ exactly equal any irrational number.

Just as it is sometimes hard to determine whether a specific number is rational, it is also sometimes difficult to determine whether a specific number is irrational. There are some irrational numbers that you are probably familiar with and many others that you are not. We can construct the first few places of some irrational numbers in several ways. We can actually create a non-terminating, non-repeating decimal that clearly has no finite sequence of digits that repeats (0.10100100010000...). Or we can find more and more digits for numbers like $\sqrt{2}$ (1.4142135623731...) by using the definition of square root and computing better and better approximations. Or we can consult resources to find more and more digits for numbers like π (3.141592653589...). Of course, we can never write down such numbers' infinite decimal expansions, but it is possible to prove that these are irrational numbers by using a deductive argument. A surprising fact about irrational numbers is that there are far more of them than rational numbers!

Continuing to build larger and larger sets of numbers, we expand to the *real numbers,* defined to be the set of rational numbers together with the irrational numbers. The real numbers complete the traditional number line, filling in all its points. Real numbers are used extensively in high school and college mathematics.

Other numbers

After repeated expansions of our number system, we ask the natural question, "Are there any other numbers beyond real numbers?" Because the real numbers complete the number line, it's hard to imagine any other numbers.

Let's explore this question by looking at different types of numbers in terms of the kinds of questions that they can answer, or more specifically, the kinds of equations that they can solve. In each case below, the type of number or numbers needed for the solution follows the equation:

- $5 + x = 9$ (counting number)
- $8 + x = 8$ (whole number)
- $12 + x = 5$ (integer)
- $10x = 17$ (rational number)
- $x^2 = 5$ (irrational numbers)
- $x^3 = 7x$ (real numbers)
- $x^2 = -3$ (???)

The last problem asks for a number that when multiplied by itself gives a product of -3. Does such a number exist? The answer is yes, but it is not a real number. We must look beyond the traditional number line for a solution. If your interest is piqued, start by doing a quick search for the terms *imaginary number* and *complex number.*

Conclusion

Rational numbers are important objects of study in their own right. They have many different uses as numbers, and they introduce a host of new mathematical ideas. This chapter has shown that they also have connections to many other aspects of the school mathematics curriculum. Moreover, we have seen that they represent a sort of middle stage in the development from simpler to more complex number systems. The next chapter shifts the focus to the challenges associated with learning, teaching, and assessing knowledge of rational numbers.

Chapter 3

Challenges: Learning, Teaching, and Assessing

Beginning at a young age, students develop an informal knowledge of fractions and decimals. In the upper elementary grades, our challenge is to build on that knowledge and help students work toward the development of rational number sense that is flexible, interconnected, and easily applied in a variety of situations.

This chapter outlines four shifts that teachers can help students to make to facilitate their development of rational number sense. These shifts in understanding are closely related to the big ideas and essential understandings that are the framework for this book.

Relating Rational Numbers to Whole Numbers

As young children begin to develop quantitative concepts, they are taught to count small sets of discrete objects. Drawing on their early experiences with counting objects, students may mistakenly think that since 5 is followed by 6 in the whole number counting sequence, there are "no numbers between 5 and 6." Using similar reasoning, older students may mistakenly believe that there are no numbers between 0.7 and 0.8, or between 0.78 and 0.79.

A challenge for teachers is to help students see rational numbers as an extension of the way in which we use whole numbers (Essential Understanding 1a). Part of that challenge is to assist students in building a foundation for ultimately understanding that there are infinitely many numbers between 8 and 9 (such as $8^4/_5$ and 8.01), or between 0.7 and 0.8 (such as 0.725 and $^3/_4$), or between any two rational numbers, as described in Essential Understanding 3b.

Essential
Understanding 1a

Rational numbers are a natural extension of the way that we use numbers.

Essential
Understanding 3b

Between any two rational numbers there are infinitely many rational numbers.

Shift 1—From unrelated system to natural extension

To understand how rational numbers build on and extend whole numbers, students need to complement their counting experiences with measurement experiences. This shift from a focus on discrete quantities to a focus on continuous quantities introduces considerably more cognitive complexity but is a key foundation for rational numbers (Heibert 1992). As we emphasized earlier, and as Dougherty and colleagues (2010) stress, measurement situations provide a natural transition from understanding whole numbers to understanding rational numbers.

When measuring, we quickly discover that whole numbers may not be precise enough to denote the amount (e.g., weight, capacity, or length) of something. For example, when reporting the amount of certain unwanted chemicals found in a city's water supply or documenting the amount of hemoglobin in a person's blood sample, measurements may need to be reported to the nearest tenth, hundredth, thousandth, or millionth of a unit.

Facilitating Shift 1

The following activity could be used to help students understand the need for rational numbers and how they are related to whole numbers. This example also provides fertile possibilities for discussing precision in measurement, which historically served as an important impetus to the development of rational numbers.

Task: Mini Olympics

Students are assigned to teams to participate in events such as those listed below. The events described can easily be conducted in limited space outdoors or in a classroom.

Shot Puff: Students toss a cotton puffball from behind a line, using a shot-put technique.

Mini Javelin: Students stand behind a line and throw a straw or cotton-tipped swab like a javelin.

Long Thump: Students press the edge of one coin (or counter) with their thumb, tiddlywinks-style, onto the edge of another coin to make the lower coin jump forward.

Discus Blast: Students use one breath of air to blow a plastic bottle cap across a long table or floor while remaining entirely behind a line.

For each event, students measure the distance with a ruler that they make by using a pencil (or another non-standard unit) to mark off units on a strip of paper, as shown:

Members of each team must discuss and agree on how to write the measurements of the distances attained in each event in pencil units (or units based on another selected item).

During the Mini Olympics activity, students will quickly find that measuring the distances in whole numbers of units is not precise enough to distinguish their measurements from one another and from other teams' measurements. They will need to subdivide the pencil units on their ruler into equal parts, such as fourths, sixths, eighths, or tenths. Being able to partition units into subdivisions other than halves, such as thirds and sixths, is an important element of rational number sense (Lamon 1999; Pothier and Swada 1983). These experiences can help students understand the relationships between fractions such as halves and thirds, halves and fourths, or thirds and sixths. For example, a perfect learning opportunity arises when students who are trying to divide a unit into thirds begin by drawing a line down the middle, find themselves faced with halves, and are then stumped! Reflect 3.1 explores ways to maximize the value of the Mini Olympics task.

tip Teaching Tip
Encourage students to partition units into equal subdivisions other than halves, such as thirds, fifths, and sixths.

Reflect 3.1

What discussion prompts related to the Mini Olympics task might help students build a foundation for understanding the following ideas:

1. Rational numbers are an extension of whole numbers.

2. There are always more numbers between any two numbers on a number line.

While observing students in the Mini Olympics task, the teacher might ask students questions such as the following to stimulate new insights or make implicit understandings more explicit:

1. What changes did you make to your ruler during the activity? Explain.

2. Why did you subdivide the units on your ruler? What are some other ways to subdivide the units on your ruler? Explain which would allow you to make a more precise measurement: thirds, sixths, or tenths of a unit.

3. What is the smallest subdivision that a unit could possibly be divided into? (Ultimately, students will understand that while it may be physically impossible for them to divide the pencil unit into millionths or billionths, it is theoretically possible to divide the unit into *any* number of equal parts.)

4. How could you convince someone that there are numbers between 2 and 3? Use your ruler to give examples of numbers between 2 and 3. (Even young students have the capacity to see that $2\,^1/_2$ is a number between 2 and 3.)

In addition to helping students build their understandings, discussion prompts such as these may be used as an informal way to assess how well students are making the transition from viewing whole numbers and rational numbers as separate and unrelated systems of numbers to understanding that rational numbers are a natural extension of whole numbers.

Multiple Interpretations of Rational Numbers and Units

A common classroom model for fractions is a circular or rectangular region that is divided into equal parts. Variations of this model are often used to describe a part-whole relationship, where a fraction is named to tell how many parts are designated (the numerator) out of the total number of parts (the denominator). One of the merits of this model is that it builds on students' life experiences with cutting things, like birthday cakes, for example, into equal shares. Another advantage is that students find it easy to use counting to name the numerator and denominator of a fraction, as when they count to name 3 parts out of 8 equal slices of the birthday cake as $^3/_8$ of the cake.

The part-whole model does have limitations, however. First, dividing a region into a specific number of parts—for example, fifths, sevenths, or tenths—is often difficult for students. Second, the model may not give students a sense of the relative magnitude of one fractional part as compared to another, and this awareness is a critical part of rational number sense. How can students understand the meaning of $^3/_4$ if they think of the 3 and 4 only as two separate numbers, each obtained by counting parts? By contrast, if they have to think about where to place $^3/_4$ on a number line, they must consider the meaning of $^3/_4$ as a single number rather than as two numbers, each with separate values. A third limitation of the part-whole model is that students often have difficulty understanding an

Big Idea 2

Rational numbers have multiple interpretations, and making sense of them depends on identifying the unit.

improper fraction, such as $^5/_4$, since having 5 parts out of a whole that has only 4 parts may seem impossible to students. A teaching challenge related to Big Idea 2 is to help students build flexibility with multiple interpretations of rational numbers. In particular, it is critical that students learn to interpret a rational number as a measure and develop their competence with the number line model (Lamon 1999; Fosnot and Dolk 2002).

Shift 2

Students need to make a transition from a simplistic part-whole model for rational numbers to a complex understanding of multiple representations, with particular attention to the measurement interpretation and the number line model.

Shift 2—From one model to many representations

As expressed in Essential Understanding 2*a*, the concept of *unit* (or *whole*) is fundamental to the interpretation of rational numbers. A unit might be continuous, or it might have discrete elements, as in a bag of marbles or a dozen eggs. Adding to the complexity of understanding the meaning of the term *unit* is the idea that any quantity of something (such as a library of 130 books, a length of 0.75 inches, or $3\,^1/_2$ dozen cookies) can be conceptualized as a unit (see the discussion of Essential Understanding 2*g* in chapter 1). Reflect 3.2 illustrates the importance of reasoning about the idea of *unit*.

Reflect 3.2

1. Consider line segment *AB* below.

   A   B

   Draw a line segment that is 1 unit long if line segment *AB* represents—

       a. $^1/_3$ unit   *b.* $^2/_3$ unit   *c.* $1^1/_2$ units

2. How many peanuts are in 1 bag if 12 peanuts represent—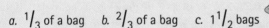

       a. $^1/_3$ of a bag   *b.* $^2/_3$ of a bag   *c.* $1^1/_2$ bags

3. Give an example where 0.6 represents more than 1 thing.

4. Give an example where $^1/_4$ of something represents exactly $3^1/_2$ things.

Solving problems like those in Reflect 3.2 requires reasoning about the unit in flexible, and sometimes complex, ways. In problem 1*a*, one might reason that if segment *AB* represents $^1/_3$ unit, then 1 unit will be 3 times as long. Or in problem 1*b*, if segment

AB is $^2/_3$ unit long, one could reason that since $^2/_3$ unit is the same as two $^1/_3$'s, a segment half as long as segment *AB* would represent $^1/_3$ unit. Then the length of this segment could be tripled to find the length of a segment 1 unit long. Similar kinds of reasoning can be used to solve the peanut problems. For example, in problem 2*c*, if 12 peanuts represent $1^1/_2$ bags, then one might think of the 12 peanuts as 3 half-bags, each with 4 peanuts. So a whole bag would have twice as many as a half-bag, or 8 peanuts.

Flexible reasoning about the unit is a key component of rational number sense, and it is something that develops over time and through many experiences. The challenge for teachers is to build on students' intuitive understanding of rational numbers to help them identify and reason about the unit.

Facilitating Shift 2

The task below sparks students' imagination about the many different kinds of things that a *whole*, or *unit*, might be. The goal is for students to expand the way in which they think about the unit to include examples such as the following:

- A unit modeled as a continuous region

- A unit modeled as a set of discrete things that are mentally "chunked" together

- A unit modeled as a single object that is composed of a number of pieces

- A unit modeled as a fractional part or a mixed number

The discussion of Big Idea 2 in chapter 1 described these ways of modeling a unit in more detail.

Task: Many Ways to Show a Rational Number

Students come up with different ways to model the fractional part $^1/_3$ (or $^3/_8$, 0.25, $2^3/_4$, or others) by drawing, using materials, or giving examples of ways in which $^1/_3$ is used in context.

 Big Idea 2

*Rational numbers
have multiple inter-
pretations, and
making sense of
them depends on
identifying
the unit.*

To stimulate discussion, and to motivate students to think in new ways, teachers ask questions such as the following:

1. Who has a way to show $^1/_3$? How do you know that your method works?

2. What about a six-pack of juice (a class of 30 students, the length of the classroom, a gallon of milk)? Could that give you an idea about another way to show $^1/_3$? Explain.

3. Give an example in which a third of something can represent more than one thing.

4. Give an example in which a third of something can be a mixed number.

Tasks such as Many Ways to Show a Rational Number are accessible to students with a wide range of prior knowledge, yet they challenge all students' thinking by bringing up situations that require increasingly sophisticated reasoning and different kinds of models. For example, it is relatively easy to illustrate $\frac{1}{3}$ of a six-pack of juice, but more challenging to illustrate $\frac{1}{4}$ of the same six-pack. Also, students may be accustomed to identifying fractional parts of a rectangle or circle but have little experience in determining the length of a line segment by using a number line model. Students benefit from thinking about the unit in a variety of real-world contexts, such as a classroom of students, a half hour, 1.2 miles, a bottle of milk, $1.50, a book, 15 homework problems, 6 granola bars, a bag of macaroni, $1\frac{3}{4}$ cups of peanut butter, $\frac{1}{2}$ yard of ribbon, or $3\frac{1}{2}$ sandwiches. Measurement contexts in particular are important for building students' rational number sense. A version of this task could also be used to assess the depth and complexity of students' understanding of rational numbers.

Learning how to show fractions and decimals with multiple models not only helps students build their rational number sense but also helps them see how these symbols are used in the real world. Developing fluency with these varied representations prepares them to think flexibly about more advanced mathematics and problems in other areas.

Understanding Rational Number Equivalence

Equivalence is an essential concept that is foundational to Big Idea 3 as well as a key component of rational number sense. Students use ideas about equivalence to compare and order rational numbers. The number line, based on the measurement interpretation of rational numbers, provides a rich context for helping students navigate rational numbers flexibly and confidently. In this context, students learn that there are rational numbers all along the number line and that there are numerous rational numbers—in fact, infinitely many—between any two given rational numbers. Students also learn to partition the unit in multiple ways so that they can find the approximate location of rational numbers on a number line. Familiarity with common benchmarks, such as 0, $\frac{1}{2}$, 0.75, and 0.5, plays an important role in understanding the relative magnitude of other rational numbers and putting them in order.

Teaching Tip

Provide students with a variety of problems that require them to identify and reason about the unit. Choose contexts that are conducive to making drawings or modeling with materials. Gradually offer problems that are more complex.

Big Idea 3

Any rational number can be represented in infinitely many equivalent symbolic forms.

Shift 3—From whole number–based to equivalence-based comparisons

When students first try to compare or order rational numbers, they attempt to make sense of the comparisons in relation to what they understand about comparing whole numbers. They may mistakenly think that $3/_5$ represents a greater value than $2/_3$ because in the numerators, 3 is greater than 2, and in the denominators, 5 is greater than 3. Or they may think (correctly) that 0.33 represents a greater value than 0.3 because (incorrect reason) 33 is greater than 3. In contrast, students can compare rational numbers in meaningful ways, based on an understanding of equivalence. Using this idea, they might reason that $2/_3$ represents a greater value than $3/_5$ because $10/_{15}$ (which is equivalent to $2/_3$) is greater than $9/_{15}$ (which is equivalent to $3/_5$). Also using the concept of equivalence, they might reason that 0.33 represents a greater value than 0.30 (which is equivalent to 0.3).

Facilitating Shift 3

The following task uses a number line to help students develop their ideas of equivalence and apply them to compare and order numbers. It also builds a foundation for the idea that there are infinitely many numbers on a number line.

Task: A Number Clothesline

Students create a number line by pinning cards, each labeled with a different rational number, to a clothesline strung across the room. Teachers should attach a few key numbers, such as endpoints or benchmarks like 0, 0.1, $1/_2$, or 1, to the number line before students begin the task. After the students begin working, if a student disagrees with someone else's placement of a number, he or she can suggest a different placement and give a justification. The discussion about each number continues until the students reach a consensus.

Sample Rational Number Sets

- Fractions with numerators of one: $1/_3$, $1/_5$, $1/_6$, $1/_8$, $1/_{10}$, $1/_{20}$

  Locate the points for 0 and 1 on the clothesline prior to beginning the task.

- Decimals and fractions: $3/_5$, $5/_{10}$, 0.35, 0.5, 0.33, $1/_3$

  (Note that 0.33 is not equal to $1/_3$ but is a close approximation of it.)

Locate the points for 0 and 1 on the clothesline prior to beginning the task.

- Fractions that include improper fractions along with mixed numbers:

$^0/_3, \ ^4/_4, \ ^5/_4, \ ^5/_3, \ ^6/_3, \ 1\,^1/_2, \ ^4/_2, \ 2\,^1/_2$

Locate the points for 0, 1, and 2 on the clothesline prior to beginning the task.

Teachers can offer this task to students repeatedly, in a single grade or across grades, with different sets of numerals. To vary the task, a teacher might change the position of 1 after all the numbers have been placed so that the numbers need to be repositioned. Or instead of starting the task by pre-locating endpoints and benchmarks such as 0 and 1, the teacher might begin by locating the points for $1\,^1/_4$ and 2, or 0.5 and 1. Students can begin to see the relationships among different forms of numbers by doing many variations of the number line task.

During an activity such as A Number Clothesline, teachers can observe areas of strength and weakness for the group as a whole as well as for individual students. They can use this information to plan other math tasks and to choose the numbers for the next clothesline lesson. They can also assess students' performance—individually, in pairs, or in small groups. An open-ended assessment task might be to ask students to choose four numbers and place them on the number line. Students could be challenged to include particular kinds of numbers, such as mixed numbers or decimals in tenths and hundredths, to allow them to demonstrate the breadth and depth of their understanding.

Teaching Tip

Provide students with assessment tasks that allow them to demonstrate ways to compare and order rational numbers by using a variety of models and symbolic forms.

Making Sense of Rational Number Operations

When rational number operations are taught as unrelated rules that are hard to remember and don't seem to make sense, students often become discouraged. Why do they add the numerators and not the denominators when adding similar fractional parts? Why do they invert the second number and not the first when dividing one fraction by another? Why do they get a smaller number rather than a larger number when multiplying two decimals like 0.5 and 0.8? When working with decimals, why can they lop off some zeros but not others, or annex zeros in some places but not others? No wonder students throw up their hands! However, by focusing on the essential understandings related to Big Idea 4, we can help students use their comprehension of whole number operations to make sense of rational number operations.

Shift 4—From "rules" to sense making

If students see meaning in addition, subtraction, multiplication, and division with whole numbers rather than just memorizing arbitrary rules, they can apply these ideas to understand meanings of these operations in fraction or decimal contexts. Although there are subtle—and sometimes complex—differences depending on the situation, the broad ideas still apply.

As in working with whole numbers (see Dougherty et al. 2010), students must become familiar with many interpretations of the operations for rational numbers. The discussion of Big Idea 4 in chapter 1 provides ideas that can assist in planning lessons to help students make sense of rational number computation in general, as well as some of the common rational number algorithms.

Facilitating Shift 4

The following task provides a setting for helping students unravel the meanings of rational number operations through experiences that connect to a real-world context. Problems situated in everyday contexts are particularly important since they provide students with opportunities to draw on contextual, linguistic, and visual cues to meaningfully interpret problems involving rational number operations.

Task: Cooking for a Class Party

Working in small groups, students are challenged to think about what an operation means in the context of measuring ingredients for recipes.

Each group of students works with a copy of a recipe that makes more or fewer servings than the number of students in the class. The students alter the recipe so that it will provide about one serving for each student in the class. For example, students might need to double or triple the recipe, or convert to $1/2$ or $1\,1/2$ times the recipe. Their job is to figure out how much of each ingredient is needed so that the new recipe will make more or less food that still tastes the same.

Each group of students should have a variety of measuring cups and spoons (they can share if complete sets are not available, or they can make drawings if no measuring implements are available).

Tasks such as Cooking for a Class Party provide an opportunity for students to develop understanding through a real-life context. If a granola bar recipe calls for $^3/_4$ cup of oatmeal and they need $1^1/_2$ times as much, they can use real measuring cups or a drawing to reason through the problem. Likewise, if a nut bread recipe calls for $^3/_4$ cup of almonds in the batter and a sprinkling of $^1/_4$ cup of almonds on top, the context can help make it obvious that they need 1 cup of almonds altogether. If students try to use computational algorithms before they develop number sense by solving problems in context, they may have difficulty in recognizing when a procedure or a solution is nonsensical. For example, when adding $^3/_4 + ^1/_4$ (without context), they may fall into the common error of adding the numerators and adding the denominators to get a sum of $^4/_8$, or $^1/_2$. This solution would not make sense to students whose rational number sense has a strong foundation gained by using reasoning to solve problems in context.

Although it is important to begin with problems situated in context, ultimately students also need to bring the same rational number sense to problems that are purely symbolic (Sfard 2000). As teachers, we not only want students to illustrate their thinking with drawings and give verbal explanations, but we also want to encourage them to express these ideas in symbolic form.

Teachers can learn a lot about students' understanding of rational number operations through discussion related to the cooking task. For example, a teacher could ask questions such as, "Suppose you had only $1^3/_4$ cups of milk and you needed 2 cups. How much more would you need? How could you write a number sentence to represent and solve this problem?" The teacher could then observe, for example, whether students think about this as a comparison problem involving subtraction ($2 - 1^3/_4 = \square$) or as a missing addend problem ($1^3/_4 + \square = 2$). In this process, the teacher could also learn how students go about calculating the result and reporting the answer. Or teachers might ask questions to assess students' handling of similar issues in the context of finding $1^1/_2$ times as much as $^3/_4$ cup of oatmeal.

Teaching Tip

In choosing problems involving operations with rational numbers, deliberately select problems that vary in complexity and context to give students experience in working with different kinds of units and representations.

Teaching Tip

Ask students who are working on problems in a real-world context to express their understanding verbally, demonstrating with materials and drawings, and symbolically, with mathematical expressions. After students have had many contextual experiences, ask them to interpret the meaning of purely symbolic problems in many ways.

Teaching Tip

If students have difficulty extracting the mathematical ideas from a rational number situation, suggest that they begin by making sense of the situation with whole numbers substituted for rational numbers. Explicitly calling attention to the parallels between problems with whole numbers and rational numbers can help students make these associations themselves.

Conclusion

This chapter has described ways to help students build on what they know about whole numbers to make sense of rational numbers. It has also discussed important shifts in understanding that occur as students build their rational number sense. Taken together, the chapters in this book are a resource for developing our own mathematical understanding of rational numbers and helping us transfer those understandings into engaging learning experiences for our students.

References

Bassarear, Tom. *Mathematics for Elementary School Teachers.* 4th ed. Boston: Houghton Mifflin, 2008.

Behr, Merlyn, Richard Lesh, Thomas R. Post, and Edward A. Silver. "Rational Number Concepts." In *Acquisitions of Mathematics Concepts and Processes,* edited by Richard Lesh and Marsha Landau, pp. 91–126. New York: Academic Press, 1983.

Bourbaki, Nicolas. *Elements of the History of Mathematics.* New York: Springer, 1998.

Carraher, David William. "Some Relations among Fractions, Ratios, and Proportions." Paper presented at the Seventh International Congress on Mathematics Education (ICME-7), Quebec, 1992.

———. "Learning about Fractions." In *Theories of Mathematical Learning,* edited by Lester P. Steffe, Pearla Nesher, Paul Cobb, Gerald A. Goldin, and Brian Greer, pp. 241–266. Mahwah, N.J.: Lawrence Erlbaum Associates, 1996.

Crawfurd, John. "On the Numerals as Evidence of the Progress of Civilization." *Transactions of the Ethnological Society of London.* Royal Anthropological Institute of Great Britain and Ireland, 1863.

Dougherty, Barbara J., Alfinio Flores, Everett Louis, and Catherine Sophian. *Developing Essential Understanding of Number and Numeration for Teaching Mathematics in Prekindergarten–Grade 2.* Essential Understanding Series. Reston, Va.: National Council of Teachers of Mathematics, 2010.

Eels, W. C. "Number Systems of the North American Indians." *The American Mathematical Monthly* 20 (November–December 1913): 263–72, 293–99.

Falkner, Karen P., Linda Levi, and Thomas P. Carpenter. "Children's Understanding of Equality: A Foundation for Algebra." *Teaching Children Mathematics* 6 (December 1999): 232–36.

Filep, László. "The Development, and the Developing of, the Concept of a Fraction." Centre for Innovation in Mathematics Teaching. *International Journal for Mathematics Teaching and Learning* (electronic), April 18, 2001. http://www.cimt.plymouth.ac.uk/journal/lffract.pdf.

Fosnot, Catherine Twomey, and Maarten Dolk. *Young Mathematicians at Work: Constructing Fractions, Decimals, and Percents.* Portsmouth, N.H.: Heinemann, 2002.

Hiebert, James. "Mathematical, Cognitive, and Instructional Analyses of Decimal Fractions." In *Analysis of Arithmetic for Mathematics Teaching,* edited by Gaea Leinhardt, Ralph Putnam, and Rosemary A. Hattrup, pp. 283–322. Hillsdale, N.J.: Lawrence Erlbaum Associates, 1992.

Kieren, Thomas E. "Rational and Fractional Numbers as Mathe-
    matical and Personal Knowledge: Implications for Curriculum
    and Instruction." In *Analysis of Arithmetic for Mathematics
    Teaching,* edited by Gaea Leinhardt, Ralph Putnam, and
    Rosemary A. Hattrup, pp. 323–71. Hillsdale, N.J.: Lawrence
    Erlbaum Associates, 1992.

Lamon, Susan. J. "Rational Numbers and Proportional Reasoning:
    Toward a Theoretical Framework for Research." In *Second
    Handbook of Research on Mathematics Teaching and Learning,*
    edited by Frank K. Lester, pp. 629–67. Charlotte, N.C.:
    Information Age; Reston, Va.: National Council of Teachers of
    Mathematics, 2007.

———. *Teaching Fractions and Ratios for Understanding.* Mahwah,
    N.J.: Lawrence Erlbaum Associates, 1999.

Lobato, Joanne, and Amy B. Ellis. *Developing Essential
    Understanding of Ratios, Proportions, and Proportional
    Reasoning for Teaching Mathematics in Grades 6–8.* Essential
    Understanding Series. Reston, Va.: National Council of Teachers
    of Mathematics, 2010.

Moss, Joan, and Robbie Case. "Developing Children's Understanding
    of the Rational Numbers: A New Model and an Experimental
    Curriculum." *Journal for Research in Mathematics Education* 30
    (March 1999): 122–47.

Murdoch, John. "Notes on Counting and Measuring among the
    Eskimo of Point Barrow." *American Anthropologist.* American
    Anthropological Association, 1890.

National Council of Teachers of Mathematics (NCTM). *Principles
    and Standards for School Mathematics.* Reston, Va.: NCTM,
    2000.

———. *Curriculum Focal Points for Prekindergarten through Grade 8
    Mathematics: A Quest for Coherence.* Reston, Va.: NCTM, 2006.

———. *Focus in High School Mathematics: Reasoning and Sense
    Making.* Reston, Va.: NCTM, 2009.

Niven, Ivan. "A Simple Proof that π is Irrational." *Bulletin of the
    American Mathematical Society* 53 (June 1947): 509.

Norton, Anderson H., and Andrea V. McCloskey. "Modeling
    Students' Mathematics Using Steffe's Fraction Schemes."
    Teaching Children Mathematics 15 (August 2008): 48–54.

Otto, Albert D., Janet Caldwell, Cheryl Ann Lubinski, and Sarah
    Hancock. *Developing Essential Understanding of Multiplication
    and Division for Teaching Mathematics in Grades 3–5.* Essential
    Understanding Series. Reston, Va.: National Council of Teachers
    of Mathematics, forthcoming.

Parker, Melanie, and Gaea Leinhardt. "Percent: A Privileged
    Proportion." *Review of Educational Research* 65 (Winter 1995):
    421–81.

Pothier, Yvonne, and Daiyo Swada. "Partitioning: The Emergence of Rational Number Ideas in Young Children." *Journal for Research in Mathematics Education* 14 (November 1983): 307–17.

Pugalee, David K., Fran Arbaugh, Jennifer M. Bay-Williams, Ann Farrell, Susann Mathews, and David Royster. *Navigating through Mathematical Connections in Grades 6–8. Principles and Standards for School Mathematics* Navigations Series. Reston, Va.: National Council of Teachers of Mathematics, 2008.

Ross, Sharon R. "Parts, Wholes, and Place Value: A Developmental View." *Arithmetic Teacher* 36 (February 1989): 47–51.

Sfard, Anna. "On Reform Movement and the Limits of Mathematical Discourse." *Mathematical Thinking and Learning* 2 (January 2000): 157–89.

Schmandt-Besserat, Denise. *The History of Counting.* Illustrated by Michael Hays. New York: Harper Collins, 1999.

Steffe, Leslie P. "A New Hypothesis Concerning Children's Fractional Knowledge." *Journal of Mathematical Behavior* 20 (2001): 267–307.

Thompson, Patrick W. "The Development of the Concept of Speed and Its Relationship to Concepts of Rate." In *The Development of Multiplicative Reasoning in the Learning of Mathematics,* edited by Guershon Harel and Jere Confrey, pp. 181–234. Albany, N.Y.: State University of New York Press, 1994.

Titles in the Essential Understanding Series

The Essential Understanding Series gives teachers the deep understanding that they need to teach challenging topics in mathematics. Students encounter such topics across the pre-K–grade 12 curriculum, and teachers who understand the related big ideas can give maximum support as students develop their own understanding and make vital connections.

Developing Essential Understanding of—

Number and Numeration for Teaching Mathematics in Prekindergarten–Grade 2
ISBN 978-0-87353-629-5   Stock No. 13492

Addition and Subtraction for Teaching Mathematics in Prekindergarten–Grade 2
ISBN 978-0-87353-664-6   Stock No. 13792

Rational Numbers for Teaching Mathematics in Grades 3–5
ISBN 978-0-87353-630-1   Stock No. 13493

Algebraic Thinking for Teaching Mathematics in Grades 3–5
ISBN 978-0-87353-668-4   Stock No. 13796

Multiplication and Division for Teaching Mathematics in Grades 3–5
ISBN 978-0-87353-667-7   Stock No. 13795

Ratios, Proportions, and Proportional Reasoning for Teaching Mathematics in Grades 6–8
ISBN 978-0-87353-622-6   Stock No. 13482

Expressions, Equations, and Functions for Teaching Mathematics in Grades 6–8
ISBN 978-0-87353-670-7   Stock No. 13798

Geometry for Teaching Mathematics in Grades 6–8
ISBN 978-0-87353-691-2   Stock No. 14122

Statistics for Teaching Mathematics in Grades 6–8
ISBN 978-0-87353-672-1   Stock No. 13800

Functions for Teaching Mathematics in Grades 9–12
ISBN 978-0-87353-623-3   Stock No. 13483

Geometry for Teaching Mathematics in Grades 9–12
ISBN 978-0-87353-692-9   Stock No. 14123

Statistics for Teaching Mathematics in Grades 9–12
ISBN: 978-0-87353-676-9   Stock No. 13804

Mathematical Reasoning for Teaching Mathematics in Prekindergarten–Grade 8
ISBN 978-0-87353-666-0   Stock No. 13794

Proof and Proving for Teaching Mathematics in Grades 9–12
ISBN 978-0-87353-675-2   Stock No. 13803

Forthcoming:
Developing Essential Understanding of—

Geometry for Teaching Mathematics in Prekindergarten–Grade 2

Geometric Shapes and Solids for Teaching Mathematics in Grades 3–5

Visit www.nctm.org/catalog for details and ordering information.